ADVANCE PRAISE FOR
REGULATING BODIES

"Rupturing the silos that too often mark scholarly discussions of sport, Jaime Schultz has brought her sharp, critical eye to the actions of sports governing bodies' elaborations of sports' eligibility rules. By exploring those decisions determining who may play, she shifts attention from the ways sport disciplines bodies to the dehumanizing processes that reduce athletes to their bodies and impose rules that determine not what those bodies may do, but which bodies may participate and on what terms. The discussion explores distinction, discrimination, and discriminatory practice in sport's eligibility rules as well as the intended, unintended, and unanticipated consequences of regulating for sex, weight, impairment, enhancement, and genetics. Through this focus, Schultz weaves together decisions and actions usually considered in isolation, identifying problematic patterns in the regulation of participation across several fields that exacerbate sport's deep seated, profound problems. Schultz's analysis produces a call to action to shift the ways sport governance and analysts approach regulation and eligibility rules that those of us with an interest in making sport better, fairer, and more just would do well to act on."

—**Malcolm MacLean** (*he/him/his*), University of Gibraltar

"In blessedly jargon-free language, Jaime Schultz persuasively shows that many of the biocultural regulations adopted by major athletic governing bodies have seriously damaged elite sports because they are ill thought out and/or empirically vacuous. However, Schultz is under no illusion that top-level sports can function without such protective regulations, which is why in the last chapter of her book she offers several thoughtful policy prescriptions for how sports authorities and their charges might produce better protective regulations that truly serve athletes and the sports they dedicate so much of their lives to. This book is a must-read for scholars in sport humanities, sport administration, and sport management, as well as those countless followers of top-level sports who take their sports seriously enough not to gloss over their all too real problems."

—**William Morgan**, University of Southern California

"Within *Regulating Bodies*–and in characteristically eloquent, insightful, and empirically grounded fashion–Jamie Schultz complicates the various policy initiatives through which elite sport has purportedly sought to protect athletes' bodies. Providing historical backdrops to what are very real contemporary issues within sport, *Regulating Bodies* offers a contextual understanding which cogently highlights the conjoined scientific and socio-cultural dimensions of sport policies which, quite literally, police as they *protect* the age, weight, sex, impairment, and natural dimensions of athletic bodies. Crucially, *Regulating Bodies* not only speaks to the biocultural derivation of protection-oriented sport organizations and politics, it also elucidates their oftentimes stark and inhumane material and social effects. As balanced and nuanced a sport studies academic as exists, Jaime Schultz utilizes various sporting case studies in presenting a compelling case which elucidates the benefits and pitfalls for implementing protective policies in elite sport. In doing so, she points to the on-going tensions/contradictions framing the present and immediate future of the domineering sport-media-commercial complex. *Regulating Bodies* represents a vital, prescient, and empirically rich contribution to the burgeoning literature focused on the socio-material dimensions of contemporary sport: it is a necessary acquisition for anyone with an interest in making sense of the socio-scientific dimensions, and compounded human outcomes, of the contemporary sporting leviathan."

—**David L. Andrews**, University of Maryland

"Combining meticulous research, unflinching critical acumen, and extraordinary empathy, Jaime Schultz's *Regulating Bodies* explores how the protective policies and regulations ostensibly meant to safeguard sport often perpetuate discrimination, danger, and abuse. Schultz takes sport to task without denying its many pleasures. *Regulating Bodies* also establishes a blueprint for how we might reinvent sport in ways that foster greater healthiness, inclusion, and enjoyment. Its observations are at once lacerating and hopeful. Anyone who cares about sport—from scholars to athletes to policymakers—should read and learn from this important and timely book."

—**Travis Vogan**, University of Iowa

"Jaime Schultz has long displayed a talent for interpreting issues in contemporary sport through the insights of social science and the perspective of history. In *Regulating Bodies*, she examines protective policies in sports, an engaging lens into the question of what it means to be an elite athlete. Clearly written, deeply researched, and framed to compel analysis and reflection, the book reinforces that Schultz is one of our smartest, most reasonable guides for how to tackle the regulatory controversies in sports."

—**Aram G. Goudsouzian**, University of Memphis

"Quite simply, *Regulating Bodies* is a remarkable book. Beautifully written and expertly researched, it exposes the complexities of some of the most pressing and controversial issues in contemporary sport. *Regulating Bodies* is a must-read for anyone interested in the past, present, and future of elite sport. A warning though, once started, it is very hard to put down!"

—**Holly Thorpe**, University of Waikato

REGULATING BODIES

ELITE SPORT POLICIES AND THEIR UNINTENDED CONSEQUENCES

JAIME SCHULTZ

OXFORD
UNIVERSITY PRESS

OXFORD
UNIVERSITY PRESS

Oxford University Press is a department of the University of Oxford. It furthers
the University's objective of excellence in research, scholarship, and education
by publishing worldwide. Oxford is a registered trade mark of Oxford University
Press in the UK and certain other countries.

Published in the United States of America by Oxford University Press
198 Madison Avenue, New York, NY 10016, United States of America.

Library of Congress Cataloging-in-Publication Data
Names: Schultz, Jaime, author.
Title: Regulating bodies : elite sport policies and their unintended
consequences / Jaime Schultz.
Description: New York, NY : Oxford University Press, [2024] |
Includes bibliographical references and index.
Identifiers: LCCN 2023057271 (print) | LCCN 2023057272 (ebook) |
ISBN 9780197616499 (hardback) | ISBN 9780197616512 (epub)
Subjects: LCSH: Professional sports—Rules—Social aspects. |
Professional sports—Safety measures—Social aspects. |
Athletes—Health and hygiene. | Sports—Physiological aspects.
Classification: LCC GV734 .S38 2024 (print) | LCC GV734 (ebook) |
DDC 796.01/9—dc23/eng/20240212
LC record available at https://lccn.loc.gov/2023057271
LC ebook record available at https://lccn.loc.gov/2023057272

DOI: 10.1093/oso/9780197616499.001.0001

Printed by Sheridan Books, Inc., United States of America

For Paul

Contents

Figures and Tables

Figures

Tables

Acknowledgments

This book took a long time to write, and I am thankful to those who have helped me along the way. I first thank those generous souls who consented to interviews: Emma Beckman, Jonna Belanger, Harvey Fink, John Gleaves, David Legg, Payoshni Mitra, Shawn Morelli, Daniel Mothley, Robin Parisotto, Matthew Porteus, Stephen Roth, Robert Steadman, and Sean Tweedy.

Several interviewees also graciously read chapters and offered invaluable feedback, as did James Druckman, Aram Goudzouzian, Thomas Hunt, Rita Liberti, Elizabeth Sharrow, and my colleagues in the History and Philosophy of Sport program at Penn State. Over the years, our biweekly meetings have brought together a wonderful cast that includes Zach Bigalke, Aaron Bonsu, Ann Cook, Mark Dyreson, Jacob Fredericks, Francisco Javier López Frías, Eli Hunter, Cam Mallett, Emmanuel Macedo, Rachel Park, Michelle Sikes, and Tom Rorke. I am also grateful for the support of my other colleagues at Penn State, especially Nancy Williams, chair of the Department of Kinesiology. Daryl Adair, Matthew Bowers, Georgia Cervin, Mark Connick, David Handelsman, Mike Huggins, David Hughes, Helen Lenskyj, Kevin Tallec Marsten, Gary Osmond, Tolga Ozyurtcu, Lindsay Parks Pieper, Murray Phillips, Steven Riess, Janice Todd, Wray Vamplew, and Kristen Worley have all offered expert advice along the way.

I am grateful to Lucy Randall, who edited the first book I published with Oxford University Press. Lucy connected me with Sarah Humphreville, another marvelous editor who has helped me immensely with *Regulating Bodies*. Thank you to Sarah, project editors

Emma Hodgdon and Lucy Harvey, production project manager Ganga Balaji, and copyeditor Anne Sanow.

In looking back through these acknowledgments, I am overcome by how lucky I am to know such a fine, lovely group of people. But I have been especially lucky when it comes to family. I appreciate my mother, Deborah, and my father, Robert, who sadly passed away before this book made it into print. I appreciate my ridiculous brothers, Rob and David, who are smarter and funnier than I could ever be. And, more than everything, I appreciate my partner, Paul, and our beautiful, wacky, wonderful children, Nella and Sylvie, who will certainly change the world. Don't say I didn't warn you.

Abbreviations

ABP	Athlete Biological Passport
BALCO	Bay Area Laboratory Co-operative
BMI	Body mass index
CAIS	Complete androgen insensitivity syndrome
CAS	Court of Arbitration for Sport
CRPD	United Nations' Convention on the Rights of Persons with Disabilities
CRISPR	Clustered Regularly Interspaced Short Palindromic Repeats
CTE	Chronic traumatic encephalopathy
DNA	Deoxyribonucleic acid
DSD	Differences of sex development
DTC	Direct-to-consumer
EPO	Erythropoietin
FIFA	Fédération Internationale de Football Association/International Association Football Federation
FIS	Fédération Internationale deSki/International Ski Federation, now International Ski and Snowboard Federation
GDR	German Democratic Republic or East Germany
HBB	Hemoglobin-Beta gene
IAAF	International Association of Athletics Federations, now World Athletics
IOC	International Olympic Committee
IPC	International Paralympic Committee
ISU	International Skating Union
MMA	Mixed martial arts
NBA	National Basketball Association
NCAA	National Collegiate Athletic Association
nmol/L	Nanomoles per liter
T	Testosterone

THG   Tetrahydrogestrinone or "The Clear"
UCI   Union Cycliste Internationale/International Cycling Union
UFC   Ultimate Fighting Championship
UN   United Nations
WADA   World Anti-Doping Agency

Introduction

Shallow Solutions, Deep Problems

It was hard to watch Kamila Valieva's individual free skate program at the 2022 Winter Olympic Games. The reigning European figure skating champion and holder of nine world records had been favored to win gold, but dogged by scandal, the pressure proved too much. The fifteen-year-old Russian stumbled and fell throughout her routine. When it was all over, Valieva waved her arm in disgust as she left the ice, and then broke into tears as she braced for the judges' scores that would place her outside of medal contention.

It was harder, still, to watch coach Eteri Tutberidze berate the skater when it was all over. It was not that Valieva was inconsolable. It was that no one tried to console her. "We were watching child abuse in real time," said Canadian coach Sandra Bezic. "And it was devastating. It was horrifying."[1] Even International Olympic Committee (IOC) president Thomas Bach was troubled by Tutberidze's "tremendous coldness," as he later characterized her behavior. "Rather than giving [Valieva] comfort, rather than to try to help her, you could feel this chilling atmosphere, this distance."[2] A Kremlin spokesperson brushed aside Bach's appraisal: "He doesn't like the harshness of our coaches, but everybody knows that the harshness of a coach in high-level sport is key for their athletes to achieve victories."[3]

It had been a turbulent Games for Valieva. Just days before the individual competition, she landed the first women's Olympic "quad"—a spectacular jump that includes four full revolutions in the air—to boost the Russian Olympic Committee to first place in the team competition. Almost immediately after came the news that Valieva had tested positive for trimetazidine, a prescription heart medication that the World Anti-Doping Agency (WADA) prohibits for its endurance-enhancing potential.[4]

Athletes who test positive for prohibited drugs typically face suspension, but with the individual competition still ahead, the Russian Anti-Doping Agency cleared Valieva to skate in the upcoming singles competition. The IOC, WADA, and the International Skating Union (ISU) appealed Russia's decision in the Court of Arbitration for Sport, the tribunal that adjudicates international sports disputes. The court's ad-hoc committee dismissed the appeal and permitted Valieva to continue her Olympic run, noting that WADA classifies athletes under the age of sixteen as "protected persons" who "can benefit from more flexible sanctioning rules."[5] As explained in the World Anti-Doping Code, protected persons "may not possess the mental capacity to understand and appreciate" the regulations.[6] If this is true, it seems fair to question whether young athletes should even be eligible for elite sport.

There were additional concerns that the adults in Valieva's life, and not Valieva herself, were responsible for her use of trimetazidine. Those concerns were well-founded, given Russia's recent history. In 2016, WADA determined that more than 1,000 Russian athletes were involved in, or benefitted from, a state-sponsored doping program that ran from 2011 to 2015. As WADA president Witold Bańka described it, "Russian authorities brazenly and illegally manipulated . . . data in an effort to cover up an institutionalized doping scheme."[7] WADA consequently banned the nation from international competition but allowed Russian athletes to represent the Russian Olympic Committee rather than Russia itself. Skeptics argue it is a distinction without difference and that the results of Valieva's drug test indicate that little has changed.

The doping controversy, compounded by coach Tutberidze's apparent lack of concern for her young charge, reinvigorated the debate over age-eligibility regulations in women's figure skating. For years, observers insisted that the age limit of fifteen was too young for senior competition: it left skaters too vulnerable to physical, psychological, and sexual abuse; too susceptible to injuries and disordered eating behaviors; and with careers that were far too short. Yet until the Valieva scandal, proposals to raise the age limit gained little traction.[8]

This time, the ISU reacted swiftly. Less than four months after the 2022 Olympic Games, the ISU Council proposed raising the age limit to seventeen, "for the sake of protecting the physical and mental health, and emotional well-being of Skaters."[9] Delegates passed the proposal by a vote of 100 to 16, with two abstentions.

Tara Lipinski was among the few insiders to openly criticize the decision. In 1997, a fourteen-year-old Lipinski became the youngest winner of the World Figure Skating Championship, which, at the time, was the age limit for senior eligibility. The following year, Lipinski became the sport's youngest Olympic champion. Her objection was not that she wanted younger skaters to have the same opportunities, but rather that the rule change does nothing about a culture that perpetually jeopardizes skaters' safety. As Lipinski posted on Instagram:

> How does this rule protect these athletes? [ISU members have] protected themselves from another future worldwide scandal that involves a 15 year old. But there will still be 15 year olds that are not eligible for senior competition that will continue to train under this same broken system. You just won't see it. It just will be buried.[10]

In her estimation, the ISU's new age limit is about protecting the federation, not about protecting its athletes.

★★★★★

This book is about protection in elite sport. More specifically, it is about protective policies that regulate athletes' bodies by biocultural categories—age, weight, sex, impairment, "natural," and "enhanced" (doped).[11] These categories are at once scientifically oriented and socially constructed. What seem to be obvious, organic, and quantifiable ways of measuring variations of human diversity are, in effect, historically and contextually specific constructs, dependent upon systems of measurement, methods of analysis, and the motivations of people who create and sustain them. What appear to be biological characteristics only manifest under the cultural conditions by which they are determined and made meaningful.

One's age depends on the way one marks time. One's weight depends upon the manner of scale. One's impairment (disability) status requires meeting socially constructed standards. Even more, competition classes based on age, weight, and impairment are demarcated by manufactured limits or "cut points" that change over time and place. So, too, do conceptions of "natural" and "doped," of what we prohibit and permit in elite sport.

Sex is also a social distinction. During the 2022 confirmation hearing of US Supreme Court justice Katanji Brown Jackson, Senator Marsha Blackburn brought up the controversy over transgender athletes' participation in sport before asking the then-nominee to "provide a definition for the word 'woman.'" Jackson responded: "I'm not a biologist." Yet even biologists concede that there is no universal definition of sex.[12] Nothing shows this better than sport. Over time, athletic administrators have tried to determine sex—specifically femaleness—by genitalia, chromosomes, genetics, and hormonal levels. Nothing has held up under scrutiny.[13] The truth is that Senator Blackburn was not looking for a biological definition of sex; she was looking for a political one, and she and her associates disguised their political machinations with the pretense of "protecting" women's sports.

What Is Protected?

Protective policies in elite sport—that is, sport at the highest international level—seem to have limited reach, as very few athletes achieve elite status.[14] Statisticians calculate that the chances of competing in the Summer Olympic Games are approximately 1 in 562,400.[15] Those are about the same odds as being struck by lightning. The opportunities get even slimmer for Paralympic hopefuls and for those who compete in winter sports. Still, elite policies have a trickle-down effect that scholars argue "may result in other levels of sporting competitions (e.g., recreation leagues, high school athletics, sports clubs, etc.) following suit."[16] More than that, protective policies give shading and contour to the meanings of contemporary sport by dictating what athletes can do to and with their bodies and, by extension, what we are willing to accept in the name of better sport.

Generally speaking, four types of protectionism characterize the policies explored in this book.[17] First, the policies are enacted to protect the spirit of fair play. Athletes are grouped into categories defined by sex, age, and weight to approximate competitive balance. In elite Para swimming, as in many Para sports (sometimes called disability sports), athletes compete in different "sport classes" based on their "degree of activity limitation." This is "to safeguard the integrity of fair competition," explains World Para Swimming.[18] Sport classes not only make contests more equitable; they encourage broader participation and spectator interest (Chapter 3).

Second, and in close relation to the spirit of fair play, are policies that create "protected classes" of athletes to be guarded against competitors with perceived biological advantages. World Athletics, for example, defends its Eligibility Regulations for the Female Classification by asserting that the women's division of track and field must be "protected" from women "with certain differences of sex development (DSD) (which means that they have the same advantages over women as men do over women)."[19] Although analysts rightly dispute both the nature and the magnitude of those advantages, they are the backbone of World Athletics' policy (Chapter 2).

Third, regulations are developed to protect athletes' health and well-being, as the ISU claims in its new age-eligibility policy. In much the same way, one of WADA's chief concerns is whether a particular substance or method "represents an actual or potential health risk" to the athlete (Chapters 4 and 5).[20] Both the International Ski Federation and American intercollegiate wrestling have likewise implemented regulations to discourage injurious weight-loss practices (Chapter 1). However, the rules only came in the wake of significant public crises—athletes were reducing their body weight to dangerous and even fatal extremes—and that damaged the sports' reputations.

Thus, the fourth and often least explicit motivation behind protective policies is to protect the image and interests of sport, as Lipinski charged in the ISU's 2022 decision to raise the age of senior eligibility. It is interesting to compare the concern for young figure skaters with the celebration of precocious skateboards just a few months earlier. The 2020 Summer Olympic Games, postponed until 2021 because of the COVID-19 pandemic, was the first to showcase skateboarding as a medal sport. Audiences marveled at the skateboarders' derring-do, especially in the women's events, which featured girls as young as twelve years old. In fact, four of the six women's medals went to athletes who were thirteen or younger, yet there was no outcry over their tender ages.

So what made skateboarders different? Most obvious were the absence of a doping scandal and the spectacle of a despotic coach dressing down her young prodigy. But audiences also seemed to be willing to set aside concerns about the dangers of skateboarding and the possible exploitation of children because the girls appeared healthy, companionable, and downright ebullient. In turn, the young skateboarders were good for both the sport and the IOC, which in recent years has sought to attract "youthful" and "urban" viewers.[21] It was only when young, broken figure skaters hurt that cause that the ISU enacted its protective age-based policy.

According to Rule 42 of the Olympic Charter, there is "no age limit for competitors in the Olympic Games other than as prescribed in the competition rules of an IF [International Federation]."[22] This means that while a federation like the ISU can set its age limit at seventeen, World Skate does not have to impose any limit at all. Consequently, observes philosopher Sarah Teetzel, "almost every Olympic event includes competitors that are officially still children and considered a vulnerable population in need of protection."[23] But where does protection end and discrimination begin?

The Question of Discrimination

When it comes to the subject of discrimination, there is a fine and somewhat wavy line between "discriminating" and "discriminatory." While the terms are semantically similar, their connotations matter. Elite sport is necessarily discriminating. That's why so few people can do it. Protective policies are also discriminating, in that they have to do with distinction—with differentiating between groups and individuals according to biocultural categories.

Yet critics argue that discriminating between ages, weights, sexes, impairments, and types of enhancement can also be discriminatory. When, in 1995, the International Gymnastics Federation raised the age of senior eligibility from fifteen to sixteen, Béla Károlyi, the sport's most decorated (if now disgraced) coach railed against "the bias and discrimination caused by an age limit."[24] In his opinion, any gymnast with "the maturity and talent to compete at this level should be here."[25] With this same logic, football manager Matt Busby once famously defended his conspicuously young Manchester United team: "If they're good enough, they're old enough."

Perhaps, but there are also important reasons for imposing age limits in sport, especially for athletes under the age of eighteen. Individuals below that age, according to the United Nations (UN) 1989 Convention on the Rights of the Child, "because of their vulnerability, need special care and protection."[26] UN executive Paulo David estimates that of the millions of kids who participate in organized sport, "20 percent are potentially at risk of different types of abuse, violence and/or exploitation; and 10 percent are victims of some form of violation of their human rights."[27] These percentages undoubtedly escalate when children are trained for elite competition—under what are essentially "worklike conditions" that would otherwise violate child labor laws.[28] Are age classifications, then, discriminating or discriminatory? Protective or prejudicial?

Elite track and field athletes have asked the Court of Arbitration for Sport to resolve similar questions concerning sex and impairment. Middle-distance runner Caster Semenya and Para long jumper Blake Leeper, for example, have argued that World Athletics' protective policies violate their right to nondiscrimination. In each instance, arbitrators sided with the federation on the grounds that it is "not a public authority, exercising state powers, but rather a private body exercising private (contractual) powers." As such, World Athletics "is not subject to human rights instruments."[29] Put differently, World Athletics is allowed to discriminate because of "sport exceptionalism," a term sociologist Helen Jefferson Lenskyj defines as "the belief that sport is unique and requires its own special laws and rules."[30] Those special laws and rules, including protective policies, can supersede local, national, continental, and international rights and accords.

Sport exceptionalism effectively holds athletes to different standards than their nonathlete peers. When the International Cycling Union first introduced anti-doping controls in 1966, Tour de France winner Jacques Anquetil called them "degrading" and asked, "Why should cyclists have to be suspected and controlled while any other free man can do what he likes and take what he likes?"[31] More than five decades later, sprinter Sha'Carri Richardson must have wondered the same thing when a positive test for cannabis—which WADA prohibits—disqualified her from the 2020 Olympic Games. The fact that Richardson ingested the drug in Oregon, where it is legal, was immaterial to her case. So was the fact that cannabis does not enhance performance in her 100-meter specialty.[32]

Sport exceptionalism leaves aggrieved athletes with little recourse. According to Article 61 of the Olympic Charter, any dispute "shall be submitted *exclusively* to the Court of Arbitration for Sport in accordance with the Code of Sports Related Arbitration."[33] As a matter of course, International Federations and National Olympic Committees follow the same system of forced arbitration. "This means," explains Lenskyj, "that athletes who challenge decisions of sports governing bodies on issues like eligibility, doping or other disciplinary grounds cannot appeal to human rights tribunals or law courts in their home countries." Sport exceptionalism therefore "disempowers" athletes from fighting against what would, in almost any other context, be considered unlawful discrimination.[34]

Unintended Consequences

Discrimination is an unintended consequence of protective policies. Indeed, any number of "unwelcome and unanticipated policy outcomes" result from efforts to protect the spirit of fair play, athletes, health and well-being, and the image and interests of sport.[35] Rules intended to assure fairness, for instance, unintentionally precipitate unfair acts—what philosophers Adam Pfleegor and Danny Rosenberg call "deceptive practices" that are "associated with cheating, lying and other forms of unethical conduct in sport."[36] Athletes knowingly break the rules to gain an advantage: they falsify their age, or exaggerate their level of impairment, or use a banned substance, or fiddle with their genetic makeup in the pursuit of sporting glory.

Protective policies can also encourage unhealthy practices. The use of weight classes in combat sports, such as boxing, wrestling, and martial arts, is to mitigate the size advantage of bigger athletes. Yet research shows that many fighters and grapplers compete at "unnatural" weights achieved through risky weight-cutting strategies. Conventional wisdom holds that they enjoy a competitive advantage over "naturally" lighter opponents, but cutting weight can have catastrophic repercussions.[37] It might also be unfair. As early as 1959, the American Medical Association's Committee on Injury in Sports declared that strategies used to cut weight were "not consistent with the spirit of sport in that they tend to defeat regulations designed to ensure fair and equitable competitions."[38] Recent observers make the same argument. The "ability to reduce weight," wrote one research team, subverts the "goal" of sport, which is "to find the best combination of technique, athleticism, and strategies."[39]

In much the same way, anti-doping policies have not stopped athletes from using performance-enhancing substances. Instead, they have turned to "more dangerous but less detectable drugs," as sociologist Ivan Waddington argues.[40] In addition, banning a drug or method can have "the effect of an advertisement," contends scholar Verner Møller, and encourage its use by both athletes and nonathletes alike.[41]

Finally, pundits maintain that protective policies hold back the progress of sport. One concern with the ISU's new age limit is that to date, no female skater over the age of seventeen has ever performed a quadruple jump. Thus, in an effort to protect young athletes, the ISU may have also spelled the "end of the quad" in women's competition.[42]

Protective policies might also hold back the progress of human potential. For nearly one hundred years, athletic performances improved at a fairly steady clip. Then, sometime just before the new millennium, researchers noticed that the top results had plateaued.[43] Scientists speculate that we may have reached the limits of human biological potential and that overcoming those limits will require new technologies, new rules, and new ways of modifying the human body, many of which are currently prohibited in elite sport.[44] We might only begin to understand the capabilities of human athleticism by altering protective policies—or by lifting them altogether.

Chapter Outline

With few exceptions, I am interested in elite, international sport regulations, particularly those crafted and enforced by the International Olympic Committee, the International Paralympic Committee, WADA, and the affiliated International Federations that govern individual sports on a global scale, such as the ISU, World Skate, and World Athletics. I limit the scope of each chapter to a select few sports to tease out when and why these policies were established, how they developed over time, their significant controversies and contestations, and, when appropriate, alternate strategies for regulating sport.

Chapter 1 begins with what were perhaps the first codified sporting regulations relative to human biology: the implementation of weight limits and categories. Weight-based rules first developed in eighteenth-century British horse racing, although the intent was to protect the horse, not its human rider. Over time, additional sports have incorporated weight considerations, primarily to enhance legitimacy, fairness, and commercial appeal. However, the policies also, albeit unintentionally, encourage rapid weight loss, or risky strategies athletes use to "make weight," as is the case in mixed martial arts. To curb the perils of cutting weight, authorities in international ski jumping and American intercollegiate wrestling have intervened with added policies that have opened debates about the balance between protectionism and autonomy.

That same balance teeters throughout Chapter 2 with the regulation of sex in women's international track and field. Since the 1920s, the federation now known as World Athletics has positioned its women's division as a "protected class" in need of safeguarding from intersex and transgender women, who are presumed to have some type of male-linked biological advantage.[45] While these policies are undeniably discriminating, the central questions are whether such discrimination is warranted, if "protecting" women sustains their second-class status in sport, and if there are alternatives to sport's two-sex organizational structure.

Proponents of one such alternative draw inspiration from the sport classes used in many Para sports. Chapter 3 considers the expansion of these classes as Para sport became increasingly political, commercial, scientific, and global in the late twentieth and early twenty-first centuries. These same developments have pushed organizers to devise fairer and more rigorous protocols for classification that have increased Para sport's prestige. Still, classification also leads to disputes over discrimination, marginalization, classifier error, and "intentional misrepresentation," where athletes exaggerate their activity limitations to compete in a favorable class. In all, these advancements and contestations offer broader commentary on the status of elite sport as they highlight both the constructed-ness of biocultural classifications and their material effects.

Chapter 4 focuses on the history of anti-doping initiatives, punctuated by a series of high-profile scandals. This includes a doping scandal at the 1998 Tour de France that led, in part, to the creation of WADA. As WADA assumed global authority and insisted on increasingly rigorous and invasive testing protocols, detractors charged the agency with over-regulation, the violation of athletes' human rights, a lack of scientific evidence to uphold its rules, impeding the progress of sport and human potential, and compelling deceptive practices that are arguably less safe than doping itself. For these and other consequences, they question the efficacy and, indeed, the necessity of protective anti-doping policies.

Chapter 5 continues the analysis of anti-doping regulations by turning specifically to gene doping, which, since 2003, WADA has included on its annually updated Prohibited List. The relatively unknown and as yet undetectable possibilities of altering one's genetic makeup to enhance performance challenge the conception of the "natural body" and the limits of human potential. I weigh those challenges alongside the relative lack of regulation concerning genetic testing for athletic talent and the potential for illness or injury to highlight important legal, moral, ethical, and practical concerns, including the (mis)use of genetic information, the possibilities of genetic discrimination, questions of autonomy and consent, an athlete's right to privacy, and how and whether to enforce anti-gene doping policies.

Ultimately, these protective policies offer shallow solutions to deep problems. I am careful to use the word "shallow" here, not "superficial," as these regulations do more than camouflage or scratch the surface of sport. On the contrary, protective policies provide some remedy to very real and significant problems. At the same time, it is important to recognize that those problems are the symptoms of commercial sport and the culture that maintains it. Treating those symptoms does not constitute a cure.

Protective policies cannot fix a "broken system," to use Lipinski's words. Rachael Denhollander, a lawyer, victim rights advocate, and former gymnast who exposed the rampant sexual violence in her sport, is similarly skeptical about the ISU's new age limit. "If you have athletes who are raised in an abusive reality, it's not like a switch is flipped when you turn 17 and suddenly your warped sense of reality and perception is eradicated," she told reporters. "Raising the age might be like extending the abusive system for a few more years."[46] The age-eligibility change does nothing to fix the systemic problems of early specialization, punishing training, persistent injuries, disordered eating practices, psychological stress, social isolation, sexual violence, and abusive coaching that have long plagued the sport, much of which violates the Conventions of the Rights of the Child.[47] It does nothing to change a scoring system that incentivizes spectacular skills that only small, young, vulnerable athletes can perform, or to curb spectators' thirst for those same spectacular skills. The age-eligibility policy is a shallow solution to the deeper problems that the ISU—and the sport-media-commercial complex more generally—are complicit in creating.[48]

★★★★★

Although it may not always be apparent in *Regulating Bodies*, I am neither anti-sport nor anti-protection. And I am mindful of critiques like that of journalist Franklin Foer, who laments that academics writing about sport "are capable of accomplishing the impossible: sucking all the pleasure and fun from the spectacle."[49] Even academics agree, such as Alan Bairner, who critiques the tendency to "focus on the negative aspects of sport" and "ignore the pleasure that sport gives to so many people as participants and spectators."[50]

Admittedly, it was a pleasure to see Kamila Valieva skate at the 2022 Winter Olympics—at least before her final difficult-to-watch performance. Who could deny the exquisite beauty, the preternatural athleticism in one of the greatest performances in the history of the sport? In her quadruple Salchow, Valieva, moving backward, launched off the ice with the inside edge of her left blade and, in the span of 0.7 seconds, ascended, elongated, and revolved elegantly four times in midair. She then landed cleanly on the outside edge of her right blade in an unfathomable mix of speed, and strength, and grace, and timing. "She defies the laws of physics," marveled BBC commentator Clare Balding.[51] Valieva then executed a quadruple toe loop with a triple-toe combination, which is just as complicated, and thrilling, and gorgeous as it sounds. "We will be talking about this moment for the next 100 years," predicted Lipinski in her capacity as NBC correspondent.[52] Valieva's performance was nothing short of sublime, and we gain nothing by arguing to the contrary.

But we also gain nothing by ignoring all that went into that performance—the training that began at age five; the cascade of specialized boarding schools she first entered at age six; the long hours spent on the ice and away from friends, family, schooling, and the fleeting wonders of childhood; the abusive coaching and caloric restriction that likely went on behind the scenes; the relentless training and the damage it will toll on her body; and the fact that even without the doping scandal, Valieva's athletic shelf life was already about to expire.

In the end, my intent is not to deny the pleasures of sport, but rather to ask: What are we willing to accept in the quest for *citius, altius, fortius?*

I

Regulating Weight

Kevin Lee nearly died. The mixed martial arts fighter needed to cut an incomprehensible twenty pounds in twenty-four hours to make weight for a 2017 Ultimate Fighting Championship bout. Lee had dropped the weight before. Repeatedly. But something went wrong this time, and he woke up the morning of the weigh-in at 161—six pounds over the 155 limit for his lightweight class. Frantic, his trainers put him in the tub and filled it with boiling water, a common tactic designed to wring the body of any last vestiges of hydration. It wasn't enough. When Lee stepped on the official scale at 10:30 a.m., it registered 156.

The Nevada State Athletic Commission, which oversaw the fight, granted Lee an additional hour to drop that last pound. He cannot recall what happened next. "I don't even know how I got [the weight] off," he admitted on a popular mixed martial arts (MMA) podcast. "I don't know where I was. Your mind just doesn't want to work when you're in those kinds of states."[1]

Official Ultimate Fighting Championship (UFC) weigh-ins take place between 9 and 11 a.m. the day before a fight, which gives athletes up to thirty-six hours to recover, replenish, and rehydrate after cutting weight. By the time Lee stepped into the octagon, he had ballooned to 183 pounds. Even so, the torture he put his body through proved insurmountable, and he submitted to his opponent in the third round. "It was what it was," Lee assessed. "I was going to make the weight, even if I had to cut my foot off or something. I said it before, it damn near killed me and I had to do what I had to do."[2]

Too many athletes are damn near killing themselves to achieve dangerously low body weight in the name of better sport.[3] Of particular concern in this chapter is the practice of "rapid weight loss," which, in the context of sport, typically occurs when athletes attempt to make weight just before a competition. Experts classify rapid weight loss as the reduction of more than 5 percent of one's body weight in less than one week.[4] Lee, who dropped 11 percent in just one day, must scoff at the definition. So must his MMA comrades, for although Lee's cut was extreme, studies show that dropping 5 to 10 percent of one's body weight in the week before a bout is a matter of course.[5]

Rapid weight loss is an unintended consequence of weight-based policies. As with the other policies explored in this book, administrators typically set weight limits and weight classes to protect the spirit of fair competition, to create protected classes of athletes, to protect athletes' health and safety, and to protect sport's interests. There are other benefits to the policies, too, including the legitimation of sport and increased participation. Fairer, more balanced contests also create what economist Simon Rottenberg called the "uncertainty of outcome" that appeals to spectators, broadcasters, oddsmakers, and gamblers.[6]

There are practical matters to consider when establishing weight-based policies. How many classes should there be? A lot of narrowly defined categories may be fairer to the competitors in each class and allow more people to participate, but they also create a greater number of contests, making an event difficult, lengthy, and more costly to stage. Fewer, broader categories will streamline the event, but athletes competing at the low and high end of the categories will be mismatched, as in Para sport (Chapter 3).

When and how often should athletes "weigh in" before a competition? MMA represents one end of the spectrum, horse racing the other, albeit with slightly different nomenclature. Jockeys typically "weigh out" no more than forty-five minutes before a race. Those who place must also weigh in immediately after to ensure they did not put on weight in the interim. This prevents them from taking in sustenance over the course of a long racing day that can include as many as six to eight rides.

But perhaps the most crucial question is what can be done to protect athletes from the perils of rapid weight loss, which, according to the American College of Sports Medicine, "may have substantial negative implications to health"?[7] Potential answers can be found in protective policies enacted in American intercollegiate wrestling and international ski jumping. Although ski jumping is not classified by weight, athletes have improved their performance through emaciation practices, as they have in other "weight-sensitive sports," where bodies act against gravity or are judged on aesthetics.[8] Ultimately, the protective policies in ski jumping and American intercollegiate wrestling may prove instructive to other elite sports, including MMA and horse racing.

The Weight of Modern Sport

Mesopotamians and Egyptians invented weighing technologies sometime around 3000 BCE.[9] Balance scales, first used by merchants to trade goods of equivalent value, compared the mass of an unknown quantity or object with a known standard. The need for precise and standardized measurements quickened during the seventeenth and eighteenth centuries, primarily due to changes brought by the European Enlightenment, imperialism, and the Industrial Revolution. In 1790, against the backdrop of the French Revolution, the National Assembly of France requested the French Academy of Sciences to "deduce an invariable standard for all the measures and all the weights."[10] The result was the metric system, which set standards for length (meters), volume (liters), and mass (kilograms).[11] It was well suited to contemporary scientific and technological innovation, and experts predicted that the metric system, "though conceived and created in

France, will belong to all the nations."[12] They almost got it right. To date, Liberia, Myanmar, and the United States remain the only metric system hold-outs.

The practice of weighing humans gained traction in the eighteenth and nineteenth centuries, amid the contemporaneous enthusiasm for scientific measurement and classification. Physicians first began regularly weighing their patients in the mid-1800s, followed by the advent of public "penny scales," and, later, personal scales to be used in the privacy of one's own home.[13] Human weight, once the fascination of subjective judgments, was quantified according to numeric standards, and those standards assumed deeper social meanings. The weight of military recruits, for example, signified both the individual's fitness for service and the general health of the nation. Insurance companies devised standardized height and weight tables to predict mortality and, by extension, the different rates they charged their customers.[14] By the early twentieth century, weight became the sword with which to cleave "cut points" between health and illness, fit and unfit, normal and abject.

Weight mattered in premodern sport, but only in the abstract. Ancient Olympic combat sports—boxing, wrestling, and *pankration* (an MMA precursor)—were called the *barea athla* (heavy events) "probably because heavier athletes tended to dominate owing to the fact that there were no weight classes or time limits," speculates historian Donald G. Kyle.[15] Both sport and measurement began to modernize around the same time in eighteenth-century Europe, particularly in England, the "birthplace of modern sport."[16] In the 1750s, notes historian Tony Collins, "the three most prominent British sports—horse racing, boxing, and cricket" began to replace "the rural sports of the past."[17] This transformation was marked by institutionalization, the codification of rules, and commercialization—all of which played important roles in the advent of weight-based regulations in two early modern sports: horse racing and boxing.[18]

Horse Racing

British flat racing—horse races that do not involve jumps—was among the first, if not the first sport, to modernize. Part of this was due to the ubiquity of the horse in contemporary life, but it also had to do with royal patronage and government support. King James I, who ruled from 1603 to 1625, was a keen rider, owner, and breeder who transformed the royal hunting ground at Newmarket, Suffolk into what would later become the "headquarters" of British horse racing.[19] Subsequent monarchs, notably James's grandson Charles II, who ruled England, Scotland, and Ireland from 1660 to 1685, affirmed horse racing as the "sport of kings." Even so, argues historian Wray Vamplew, "to suggest that racing was the prerogative of the élite is wrong: it was the sport of all, a common interest of peer and peasant, of lord and laborer."[20]

The weight each horse carries in a race matters tremendously—more weight slows it down; less weight frees it up. And so in the 1700s, racing officials began to standardize horses' weight allotments. One manifestation of this was "handicap" races, in which an umpire—later formalized into the position of handicapper—decided the weight each horse would carry. This included the rider, saddle, and any added allowances based on "the merits of the horses, for the purpose of equalizing their chances of winning," wrote James Rice in his 1879 *History of the British Turf* (Figure 1.1).[21] Based on their allotments, horses were referred to as "heavy weights," "light weights," and "feather weights," terms later adopted for classes in other sports.[22] As Admiral John Rous, a steward of the Jockey Club and the sport's first public handicapper explained in 1866, "A handicap is intended to encourage bad horses, and to put them on a par with the best. It is a racing lottery—a vehicle for gambling on an extensive scale, producing the largest field of horses at the smallest expense."[23] Indeed, argues Vamplew, gambling "was one of the original influences on the formulation of rules" in modern sport.[24]

Figure 1.1 Fores's Racing Envelope, c. 1840s. Print by J. R. Jobbins. Notice the jockey being weighed on the balance scale in the bottom-left corner. British Museum.

British racetrack officials also established "scales of weight" relative to the age of the horse. In 1740, out of concern for the sport's rampant and unregulated proliferation, Parliament passed "An act to restrain and prevent the Excessive Increase of Horse Races." Included in the act was the "first Weight-for-Age Scale, of any kind, that ever found its way into print," writes Irene McCanliss in *Weight on the Thoroughbred Horse.*[25] In its initial formulation, the act decreed that a five-year-old horse carried 10 stone (140 pounds); a six-year-old, 11 stone (154 pounds); and a seven-year-old, 12 stone (168 pounds). Any horse carrying less than the appropriated weight was disqualified and his owner was fined £200.[26] The act, and the hefty sum levied at underweight offenders, helped to preserve the sport for society's upper crust.

The introduction of thoroughbred horses in the early 1700s augmented the sport's elitism. The progeny of domestic mares and imported Arabian stallions, the sleek, slender-legged thoroughbred was astonishingly fast, incredibly valuable, and performed well at a younger age than other horses (Figure 1.2). In the *Racing Calendar*, first published in 1727, racing enthusiasts could read the "details of performances of all horses running in important races," including their ages and the weights they carried.[27] The *Calendar* marked "the turning-point between ancient, and what must be considered comparatively modern turf history," observes C. M. Prior, an early historian of the sport.[28] It also marked "the beginnings of what would become known as sports statistics," as Collins contends.[29]

Sometime around 1750, a group of "Noblemen and Gentlemen" formed the Jockey Club, a private social group that organized races at Newmarket.[30] Among other accomplishments, the Jockey Club advertised a 1752 "dash" (one lap around the track) and set the weight at

Figure 1.2 *Quiz, After His Last Race at Newmarket*, Henry Bernard Chalon, 1807. Yale Center for British Art, Paul Mellon Collection. Wikimedia Commons.

"eight Stone, seven Pound" (119 pounds), which remained the accepted standard weight at Newmarket until the mid-1800s.[31] Eventually, the British government ceded control of horse racing to the Jockey Club, and by 1850, the club's rules were institutionalized at all British race-tracks, including the compulsory weigh-in after a race.[32] So, too, was Admiral Rous's weight-for-age scale, which he devised according to the distance of the race and the month it took place. Although Rous's scale has since evolved and varies from one racetrack to another, its general principles remain today. Ultimately, the scale benefits specta-tors, bettors, and the horse, but it also requires that the average jockey maintain a strict and often unhealthy body weight.[33]

Boxing

Boxing likely took its cues from horse racing. Once again, gambling "lay behind the decision, in 1746, to divide boxers into light-, middle- and heavyweight classes (along the lines of horse handicapping)," pos-its scholar Kasia Boddy.[34] Yet historian Elliott J. Gorn describes an 1842 ill-fated bout as taking place in the "era before weight classes."[35] It was also the era of bare-knuckle prizefighting, replete with wrest-ling holds, spiked shoes, and an indeterminant time and number of rounds, all of which resulted in bloody, violent contests. Adding to the sport's ill repute was its association with gambling, drinking, ca-rousing, and corruption that grated against Victorian sensibilities and bourgeois values. In 1861, Britain outlawed professional prize fighting, and by 1880, it was illegal almost everywhere in the United States. Of course, none of this put an end to the sport. Men—and more than a few women—simply set out to thrash one another in shady saloons, back alleys, remote towns, isolated barns, wooded clearings, and as they drifted atop river barges.[36]

The Marquis of Queensberry Rules, drafted in London in 1865 and published in 1867, gradually rehabilitated boxing's image. The new regulations called for three-minute rounds with one-minute rests in between and ten-second knockouts. The rules also banned wrestling holds and instituted the use of boxing gloves to make the sport more "civilized." The new "gloved contests" also gave organizers a way to circumvent local ordinances against "fist fights."[37] With the British Empire's growing influence, the Queensberry Rules, as with the rules of other British pastimes, spread around the globe.

Originally intended for amateur fights, Queensberry Rules said nothing of weight classes, although boxers competed for trophies at light, middle, and heavyweight as a "gesture towards more sportsman-like competition," writes scholar W. Russel Gray.[38] It was not until the 1880s, after Queensberry's widespread adoption, that "real effort was made to standardize weight divisions both in Britain and the United States," argues sociologist Kenneth G. Sheard. "This innovation, of course, allowed boxing skill to have a greater impact upon the outcome of a contest than extra poundage or extra reach."[39]

Richard Kyle Fox, who in 1875 purchased the weekly magazine *The National Police Gazette*, played a pivotal role in popularizing prize fighting in the United States. Not content to merely report on the fights, Fox organized and promoted competitions to whet appetites and build his audience. In doing so, he turned the moribund *Gazette* into the "Bible of the Barbershop," a veritable staple anywhere men cared to congregate. Fox recognized six weight classifications in boxing—flyweight, bantamweight, lightweight, welterweight, middleweight, and heavyweight—and awarded the winner of each class the title "Champion of the World," along with a lavish "Police Gazette" belt (Figure 1.3).[40] Athletic clubs and organizations eventually standardized the use of weight classes and, in the process, breathed an air of respectability into the once-furtive sport.[41]

Still, notes historian Benjamin Rader, "Without a national regulatory body or a rational system for determining champions, boxing was an unusually chaotic and disorderly sport."[42] With the modernization of sport came bureaucratic organizations to establish rules and systematize weight categories. The National Sporting Club of London defined seven weight classes in 1904. New York's 1920 Walker Law legalized boxing in the state and empowered the New York

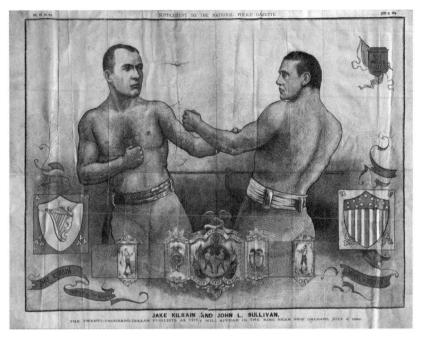

Figure 1.3 "Jake Kilrain and John L. Sullivan. The Twenty-Thousand-Dollar Pugilists as They Will Appear in the Ring Near New Orleans, July 8, 1889," *National Police Gazette.* Note that the location of the fight is only identified as "near New Orleans," since the location of the quasi-legal fight was not announced in advance. Ticket holders were taken to the fight on special trains leaving New Orleans to a secret destination. Wikimedia Commons.

State Athletic Commission to set the rules and sanction fights within the state. The commission's "adoption of definitive weight divisions," writes legal scholar Jack Anderson, was "quickly implemented throughout the United States."[43] Over time, notes political scientist Robert Rodriguez, "boxing's governing bodies came to realize that they could make more money sanctioning fights in more weight classes, the number of divisions has grown exponentially."[44]

In short, weight classifications were crucial to the standardization, legitimization, and commercialization of horse racing and, later, boxing. In the late 1800s and early 1900s, rowing, wrestling, and weightlifting adopted similar regulations. At the same time, these characteristics set in motion the problems associated with "making weight."

Making Weight in Mixed Martial Arts

In many ways, and abridged form, the history of mixed martial arts resembles the history of boxing. Linked to the *pankration* of ancient Greece and Brazilian *vale tudo* (anything goes) matches, the first modern MMA competition took place in 1993 under the auspices of the UFC, now the sport's largest promotion company. Initially, there were no weight categories. Rather, the objective was to identify the most effective fighting discipline. Pitted against each other were athletes from boxing, Brazilian jiu-jitsu, judo, karate, kickboxing, Muay Thai, sambo, Taekwondo, and wrestling. This led to some interesting matchups, such as the 1994 bout between the 6 feet 8 inch, 616-pound sumo wrestler Emmanuel Yarbrough and 5 feet 11 inch, 200-pound martial artist Keith Hackney, whose victory earned him the nickname "The Giant Killer." If there's one thing sports fans like more than an uncertain outcome, it's a David and Goliath story.

Few rules governed early MMA contests. Referees had no authority to stop the fights and vicious maneuvers such as biting, eye-gouging, and fish-hooking were fineable offenses, but not banned outright.[45] There were no time limits and victory was decided only by knockout, submission, or abandonment.[46] Predictably, critics condemned MMA, just as they had with boxing a century before. US Senator John McCain notoriously referred to it as "human cockfighting" and led a congressional effort to ban the sport. Thirty-six states complied. Cable television channels stopped broadcasting the lucrative pay-per-view events as MMA organizers scrambled to save their enterprise.

Part of their success lay, as it did in boxing, with the introduction of weight classes, first implemented in 1997 with UCF 12 (events are sequentially numbered). New UFC management further banned some of the more egregious tactics, granted the referee more authority, and added round and time limits, as in the Queensberry boxing rules of 1867. In fact, in the US, MMA came under regulation by the very organizations that governed boxing. By 2007, even McCain had dropped his crusade against the sport and conceded that it had made "sig-nificant progress."[47] Two years later, international regulatory bodies adopted the official Unified Rules of Mixed Martial Arts, and the sport is now broadcast in over 129 countries and 800 million house-holds worldwide.[48]

But regulation brought with it a series of unintended consequences, including those associated with cutting weight. Approximately 60 to 90 percent of athletes in all combat sports cut weight, but studies find that mixed martial artists exhibit the most extreme pre–weigh-in weight loss practices.[49] Researchers surveyed British MMA fighters in the flyweight to welterweight categories (125–170 pounds; see Table 1.1) and found that "One hundred per cent of MMA athletes engaged in complete fasting or low carbohydrate diets in the final 3–5 days prior to weigh-in."[50] That's every single fighter they talked to. In ex-treme cases this has led to fatalities, such as Leandro Souza who died trying to lose 20 percent of his body mass—thirty-three pounds—in the seven days before a fight. And while not all of the fighters suffer ill

Table 1.1 Ultimate Fighting Championship Weight Classes

UFC Weight Class	Upper Weight Limit, Men	Upper Weight Limit, Women
Strawweight	115 pounds	115 pounds
Flyweight	125 pounds	125 pounds
Bantamweight	135 pounds	135 pounds
Featherweight	145 pounds	145 pounds
Lightweight	155 pounds	N/A
Welterweight	170 pounds	N/A
Middleweight	185 pounds	N/A
Light Heavyweight	205 pounds	N/A
Heavyweight	265 pounds	N/A
Super Heavyweight	N/A	N/A

Source: https://www.ufc.com/news/understanding-ufc-weight-classes-and-weigh-ins

effects from cutting weight, the universality of the practice indicates that few compete at their "natural" weight.

Part of the problem is in the way MMA is organized. Fighters are independent contractors, not employees of an organization. They only get paid if they fight and are frequently offered bouts without much advanced notification, leaving them desperately scurrying to meet weight-class standards. This exacerbates the hazards of a brutal sport. As fighter Cortney Casey explained to media scholar Jennifer McClearen:

> We put a lot of wear and tear on our bodies without the proper recovery, with all the weight cutting. . . . Then add the stress put on us having to fight short notice sometimes because that's our only option. Then we're cutting drastic amounts of weight. Then we get more hurt outside of practice.[51]

The sport's biggest stars, those who rake in millions of dollars per fight in careers that stretch over a decade, are the outliers. The average career length for an MMA fighter is just 1.5 years and lasts a mere three bouts.[52]

Even during that short time, fighters go to great, often perilous lengths to make weight. This can include fasting, reduced carbohydrate and fat intake, forced vomiting, increased exercise, and the use of laxatives, diet pills, and emetics. Dehydration is both the "easiest" and the most "dangerous" way to quickly drop weight, principally because the human body is about 65 percent water.[53] As in many weight-sensitive sports, "drying out" is accomplished through fluid deprivation, the use of diuretics, saunas, or "sweat boxes," exercising in a heated room, exercising while wearing a vapor-impermeable suit (typically made of plastic or rubber suit), or "water loading," where athletes reduce their sodium intake and overdrink water to trigger a "flushing mode" and induce excessive urine production.[54] And despite the recovery time after the weigh-in, over 40 percent of MMA combatants enter the octagon significantly or seriously dehydrated.[55]

Dehydration can impair muscular performance, decrease blood plasma, increase blood viscosity, affect cardiovascular functions, and compromise kidney function. It also inhibits the sweating process, which can affect temperature regulation and, when combined with heat stress, bring on hyperthermia, a severe overheating of the body linked to the deaths of athletes in several sports. What is more, dehydration decreases cerebral spinal fluid, which increases the likelihood of traumatic brain injury in a sport like MMA, where the objective is to pummel one's opponent into submission.[56]

Women MMA fighters face additional health risks, made worse by the few weight classes available to them. The UFC did not allow women to fight until 2013, at which point it sponsored only a bantamweight class for those weighing between 126 and 135 pounds, primarily to capitalize on the star power of Ronda Rousey. As of 2024, there are just four UFC weight classes for women, while there are ten for men. The heaviest class for men is the unlimited super heavyweight division, while the heaviest class for women is featherweight, capped at 145 pounds. This is the same upper limit for women competing in Bellator and the Invicta Fighting Championships, an all-women's promotional company established in 2012. For someone like champion "Cris Cyborg" (Cristiane Justino) who typically weighs between 170 and 185 pounds, this requires making "agonizing" cuts (Figure 1.4). Before her 2016 fight against Lina Lansberg, Cyborg dropped an inconceivable twenty-four pounds in just four days.[57]

There is the stink of sexism here. MMA spectators have a hard time wrapping their heads around the idea of watching two women fight one another. Smaller, more "feminine" women make the spectacle more palatable, regardless of the processes they use to cut weight. That same stink (among others) wafts through horse racing tracks, where chauvinism, misogyny, and wrongheaded ideas about female physicality continue to hold back women jockeys.[58]

Figure 1.4 Strikeforce Weigh In, Gina Carano (left) and "Chris Cyborg" (right), 2009. Wikimedia Commons.

Horse Racing's Struggle of Weight

The scale of weights, first devised by Admiral Rous, keeps jockeys remarkably slight. American jockeys typically carry between 112 and 126 pounds, which includes both the rider, whose weight averages 110 pounds, and approximately 7 pounds of gear, including silks, pads, goggles, boots, helmet, girth, saddlecloth, and saddle.[59] If the jockey and gear come in underweight for a particular horse's assignment, that horse carries added weights to make up the difference. If the jockey and gear are overweight, that jockey is scratched from the race and forfeits any money they might have earned. It stands to reason that a horse carrying a lighter load will be able to run faster. It is also safer for the animal, but it is inherently unsafe for the riders.

Insiders estimate that 80 to 90 percent of jockeys take daily pains to strip weight. It is a sport where "losing five to 10 pounds every day is just part of the routine," observes racing insider Scott Gruender.[60] In addition to the reducing practices prevalent in MMA fighters, jockeys have also used thyroid medications, amphetamines, and cocaine.[61] And anywhere from 10 to 60 percent of jockeys regularly engage in "flipping," or forced vomiting. Older racetracks bear the mark of this practice with bathroom stalls devoted to a "heaving" or "flipping bowl"—a square porcelain basin that stands higher than a standard toilet so that jockeys do not have to kneel to purge. With or without these bowls, the practice continues. "Whether it's a toilet or a heaving bowl," remarks professional jockey Darrell Haire, flipping is "still there."[62]

Racing a thoroughbred horse is among the most dangerous and under-appreciated skills in all of sport. Perched precariously atop 2,000-pound animals that tear around a track at speeds nearing forty-five miles per hour, jockeys balance their weight on the toeholds of the thin metal stirrups. Their hands clutch the reins and a fistful of mane as they navigate through a thundering herd of up to twenty competitors. Imagine doing that weakened, dizzy, and with blurred vision and a slowed reaction time.

Now imagine doing that up to six to eight races a day without the chance for much-needed hydration and nourishment due to the weigh-out/weigh-in procedures. Jockeys, like MMA fighters, are independent contractors employed by racehorse owners, and their livelihood depends on how often they ride. Many riders race almost every day of their careers, with no defined "off-season" during which to recover.[63] It is unsurprising, then, but no less disturbing, that the average jockey's career lasts somewhere between just two and ten years, cut short by persistent injuries and the unrelenting cycle of making weight.[64]

The health ramifications are long-lasting: blood disorders, kidney and liver damage, heart and gastrointestinal tract problems, and electrolyte imbalances. "Nothing else compares to the struggle of weight," shares retired rider Jeff Bloom. "It takes such a toll on our body for life. I had to reduce 7–9 pounds most days and now I suffer from kidney stones. It's such a tough grind."[65] Bones, long denied calcium, become brittle and ache. Teeth, stripped of enamel after years of flipping, deteriorate and decay, causing pain, tooth loss, infection, and gum disease. There are also psychological ramifications to a life spent trying to make weight, including mood swings, depression, body dysmorphia, and the strain on social and familial relationships. "You can't rely on the jockeys themselves to solve the problem," assesses physician David Baron, "I think the jockeys probably need more help from within the industry."[66]

Jockeys need more help from outside the industry, too. Throughout history, there has been significant and justified outcry over the maltreatment of racehorses—doping, abuse, and animal cruelty, to name a few offenses. Yet there is comparatively little concern for their riders and the ravages of rapid weight loss.

Crises and Interventions

Mixed martial arts and horse racing could learn something from wrestling, a sport traditionally riddled with risky eating attitudes and behaviors. While this is true around the world and within all wrestling styles, the National Collegiate Athletic Association's (NCAA) adoption of stricter regulations in American intercollegiate folk style wrestling is arguably the most comprehensive approach to addressing rapid weight loss.

There were no weight classes in the wrestling competitions of the first modern Olympic Games of 1896, perhaps in tribute to the ancient heavy events. In the United States, however, the National Association of Amateur Athletes restricted entry into its inaugural 1887 tournament to wrestlers who weighed less than 135 pounds. This was not about protecting light-weight athletes, but rather about protecting amateurism. Professional wrestlers tended to be heavier, and the weight limit was to exclude their participation.[67] The Amateur Athletic Union took control of the tournament in 1888, where wrestlers competed in one of two classes: under 120 pounds and under 158 pounds. As both the sport and the number of weight classes continued to grow, there were musings on the "effects of training down," as one physician put it in 1917.[68] For the next several decades, however, the general consensus was that "cutting weight" produced "no deleterious effects."[69]

It was not until the 1950s and 1960s that "the gaunt and desiccated wrestler" became cause for concern.[70] Part of this was due to the rise of sport science and sport medicine in the postwar era, particularly concerning dehydration. Throughout most of the twentieth century, sporting mythology held that hydration was detrimental to athletic performance, even in the most arduous events. Marathon runners believed that drinking water would slow them down.[71] Cyclists training for the Tour de France were counseled to "avoid drinking when racing, especially in hot weather. Drink as little as possible . . . When you drink too much you will perspire, and you will lose your strength."[72] In hypermasculine sports like American football, drinking water was considered a sign of weakness.[73]

In the 1960s, new knowledge about hydration, coupled with the skyrocketing popularity of interscholastic wrestling, brought attention to the problem of cutting weight.[74] In 1967, the American Medical Association issued a position statement against "the hazards of indiscriminate and excessive weight reduction" and urged greater education for coaches, athletes, and their parents.[75] That same year, the local medical society in the wrestling hotbed of Muscatine, Iowa, recommended that area high schools drop the sport from their athletic programs. In the members' opinion, "many youngsters go on strict diets to enable them to slip within standards for certain weight classes."[76]

All of this prompted more serious study of weight-management strategies in wrestling, along with guidelines and recommendations from physicians and athletic trainers.[77] They had little effect. Before the NCAA intervened in the late 1990s, "weight cutting was like the wild, wild West," describes Daniel Monthley, an athletic trainer who has worked with Pennsylvania State University's championship wrestling team for over thirty years.[78] As coach Rob Prebish details, "the rubber suit was simply a much needed part of the wrestler's wardrobe. Extra workout sessions in searing heat were the norm. Starving before weigh-ins was simply a test of intestinal fortitude. If you did not cut a lot of weight for wrestling, you were not tough enough."[79] In 1990 researchers found that "little had changed from the 1940s." In their estimation, weight-loss cycles remained "frequent, rapid, and large," and proposed that "a more aggressive approach may be needed to eradicate weight control practices used by wrestlers."[80]

It took a series of tragedies to initiate such an approach. In 1997, in the span of just thirty-five days, three college wrestlers, competing for three different schools, died while cutting weight.[81] In the period preceding their deaths, the men, aged nineteen to twenty-two, had severely restricted their fluid and food intake. They had been working out in hot conditions wearing vapor-impermeable suits, trying to suck out every last drop of water from their bodies. As hyperthermia set in, they collapsed and went into cardiorespiratory arrest. Their deaths were entirely preventable.

The NCAA subsequently made a series of sweeping changes that have reduced, though not eliminated the harms associated with rapid weight loss. Perhaps the biggest change was the weight certification program that sets the lowest allowable weight at which a wrestler can safely compete. Before the season begins, a physician, athletic trainer, or registered dietician assesses a wrestler's hydration status, body fat, and weight. That information is entered into an online optimal performance calculator to "certify" the wrestler's minimum weight—the wrestler must compete in a weight class above that minimum weight for the duration of the season.[82]

In addition to the minimum weight certification, intercollegiate wrestling made other changes, including:

1. *Weight class changes:* The NCAA added 7 pounds to each weight category, such that the original 118-pound class shifted to 125 pounds and so on (Table 1.2).

2. *Shortened weigh-in schedule:* Wrestlers previously weighed in the day before tournaments and up to five hours before dual meets, which allowed for significant time to rehydrate and recover after rapid weight loss. Under the new rules, weigh-ins take place one hour before the competition starts, with allowances made for multiday tournaments. "That rule on its own really

Table 1.2 National Collegiate
Wrestling Association Weight Classes

Men	Women
125	101
133	109
141	116
149	123
157	130
165	136
174	143
184	155
197	170
183–285	191

Note: NCAA classes differ from Olympic and
international weight classes.

curtailed some of the extreme weight cutting," Monthley
explains.

3. *Competition order*: Under the old rules, meets began with
   the competition for the lightest class and progressed to
   the heavyweight, allowing bigger wrestlers more time to
   recover from the effects of rapid weight loss. Under the new
   procedures, the beginning weight class is randomly selected.

4. *Prohibited methods*: Wrestlers are prohibited from using laxatives,
   emetics, self-induced vomiting, hot boxes, steam rooms, and
   saunas. They are also prohibited from using vapor-impermeable
   suits made of rubber, rubberized nylon, or neoprene, and the
   temperature of practice facilities cannot exceed 80 degrees.

Since 1997, no NCAA wrestlers have died as a direct result of weight-
cutting practices and researchers are "cautiously optimistic."[83] One
survey found that 40 percent of college wrestlers reported that the
new rules positively influenced their weight-management behaviors.
But that also means the majority of wrestlers do not feel the same.

And studies find that dangerous practices, such as the use of saunas and rubber or plastic suits, continue despite the NCAA's ban.[84] "It comes down to the integrity of the university," says Monthley. For that reason, he recommends NCAA officials or school athletic directors implement "spot checks" for compliance. "I'm not saying the system is perfect," he continues, "but it puts the checks and balances in place and if we're listening to the checks and balances and doing things the right way, we shouldn't run into problems."

Ski jumping authorities instituted their own system of checks and balances when it became clear that athletes were reducing to unhealthy weights under the maxim "fat don't fly."[85] The trend quickened in the 1990s with the growing popularity of the "V" technique, in which skiers soared off the ramp with the back end of their skis held together and the tips splayed out in front in a V, as opposed to the earlier practice of holding the skis parallel to one another (Figures 1.5 and 1.6). The V, combined with new aerodynamic suits and dangerously low body weight, allowed athletes to sail unprecedented distances (see Figures 1.5 and 1.6).[86] "Once the province of squat-legged power jumpers," reported the *Washington Post* in 2002, "ski jumping has become a sport of featherweights, an event in which you are at least on the medal podium what you eat."[87] As Norway's Oevind Berg, the 1993 world champion told a reporter, "I jumped one meter [3.3 feet] farther for each kilogram [2.2 pounds] I lost."[88] Later research determined that Berg's calculation was exactly right.[89]

By the turn of the twenty-first century, emaciation practices in ski jumping had reached epidemic proportions. In 2004, the International Ski Federation (FIS) instituted a rule that tied a jumper's body mass index (BMI) to the maximum length of *his* skis (the organization was just on the cusp of finally offering women's competitions). The BMI is calculated by a person's body mass divided by height squared (kilograms/meter^2). The maximum ski length is 145 percent of a jumper's height. Because longer skis translate to longer jumps, the FIS maintains that a jumper's BMI must be at least a "healthy" 21 to use the

Figure 1.5 Ski Jumping, 1905. Photograph: Notman and Son, Library and Archives Canada.

Figure 1.6 Martin Koch, World Cup Ski Flying, Vikersund, 2011. Photographer: Geir A. Granviken. Wikimedia Commons.

longest allowable skis. For each BMI half unit below 21, jumpers must reduce the length of their skis by 0.5 percent.[90]

A year after the BMI regulations were implemented, world-class ski jumpers gained an average of eight pounds, prompting federation

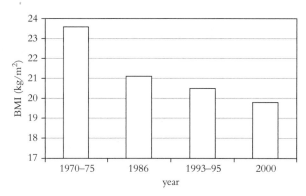

Figure 1.7 Decrease of Relative Body Weight of Ski Jumpers from 1973 to 2000. In Wolfram Müller et al., "Underweight in Ski Jumping: The Solution of the Problem," *International Journal of Sports Medicine* 27, no. 11 (2006): 926–934.

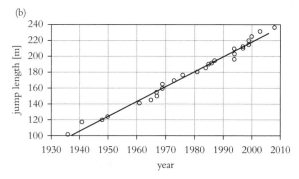

Figure 1.8 Ski Jumping (Ski Flying) World Records. In Wolfram Müller et al., "Underweight in Ski Jumping: The Solution of the Problem," *International Journal of Sports Medicine* 27, no. 11 (2006): 926–934.

officials to crow that "the underweight problem in ski jumping has almost disappeared."[91] Other assessments are not as sanguine. While reviewers concede that the "FIS succeeded in stopping the alarming development of underweight problems in ski jumping," they also find that athletes are willing to drop below the 21 BMI threshold. Being underweight makes up for the shorter skis.[92] It may be worth the penalty to drop a few more pounds.

Balancing the Scales?

So what is to be done? As long as weight limits and weight classes exist—and they exist for good reasons—athletes will manipulate their body weight. Certain sports, such as NCAA wrestling and international ski jumping, have made some headway in curbing the health risks associated with rapid weight loss. Even so, dangerous practices persist. And they would persist even if organizations explicitly banned rapid weight loss from sport. The NCAA bans the use of rubber suits, but wrestlers still wear them to cut weight. WADA prohibits the use of diuretics, yet the use of those drugs constitutes the majority of doping offenses in combat sports.[93]

Educational initiatives also have limited effect. Athletes typically turn to coaches and peers for weight-management advice. Intuitively, it would seem that making sure these parties have the best health and safety information will curb the dissemination of dangerous advice, but research also shows that just isn't the case. Some sports are simply too mired in tradition and folklore for education to affect deeper problems.[94]

Organizational suggestions include changing the existing weight classes and, in horse racing, the scale of weights to better reflect secular trends. Adding more weight classes might also help. Some pundits recommend reclassifying combat athletes by height. Because adult height is unchangeable and associated with limb length, proponents of the plan contend that it is a more equitable and healthy alternative to weight classes.[95]

Procedurally, weighing in just before a competition might discourage rapid weight loss. Another option is to set a maximum weight regain allowance that would keep athletes nearer to their certification weight, although this hasn't really helped in horse racing. Additionally, critics worry that athletes will continue to cut weight and, without adequate recovery time, face further health risks. In that regard, hydration tests might be used as part of the weigh-in procedures, where weight is only certified if the athlete is adequately hydrated. But this requires time, money, expertise, and surveillance that many sports do not care to implement or cannot afford.

A lack of both resources and concern makes other changes unlikely, such as greater oversight from independent physicians, nutritionists, dieticians, and other health care specialists, who might do more to monitor athletes' well-being, limit the amount of weight lost during a specific time period, and set minimum competition weights. This can be especially difficult in sports where athletes are independent contractors, lack the collective bargaining power of a union, or where they are without the protections accorded by a unifying governing body.

"The truth is that we are forced to view the issue from a libertarian perspective," reasons journalist James MacDonald. "We must respect the athletes' right to do as they please with their own body so long as they are not harming anyone else."[96] Still, when public weigh-ins are part of the sporting spectacle, when individuals are valorized for prodigious weight loss, and when dangerous weight loss is seen as part of the sporting ethos, establishing protective weight-based policies seems to be a shallow solution.

2

Regulating Sex

The 2020 Tokyo Summer Olympic and Paralympic Games were unique in many respects: they didn't actually take place in 2020; they happened at the height of the global COVID-19 pandemic; Japan was under a state of emergency due to rising coronavirus infections and new variants; and no spectators were allowed to attend. Within the context of so many unprecedented events, it was easy to miss another important milestone in Olympic history—the participation of the first out transgender athletes.

The IOC initially addressed the issue of transgender inclusion in 2003, but out participation was something new in 2021. This included New Zealand weightlifter Laurel Hubbard.[1] Although critics protested her presence, Hubbard, a transgender woman, met the International Weightlifting Federation's eligibility criteria at that time: to compete in women's weightlifting events, an athlete had to register less than 10 nanomoles of testosterone per liter of serum (10 nmol/L)—a specific measure of the naturally produced sex hormone present in one's blood. Hubbard's testosterone levels, tests confirmed, were below that limit and had been for more than twelve months, as the policy required.[2]

With respect to women's testosterone, the Games were also noteworthy for who was *not* there. Conspicuously absent were Namibia's Beatrice Masilingi, a specialist in the 400-meter race, Caster Semenya of South Africa, the reigning 800-meter Olympic champion, and Margaret Wambui of Kenya, who took bronze behind Semenya in

2016. World Athletics barred their participation after determining that the women, who were assigned female at birth, naturally produce too much testosterone. What is more (or less), the threshold for women track athletes was half that of weightlifters (less than 5 nanomoles of testosterone per liter of serum [5 nmol/L]). Female athletes who registered above this limit had to suppress their endogenous (natural) testosterone for a period of six months to be eligible to compete as women.

Strangely, World Athletics' regulations only applied to women who competed in distances between 400 meters and 1 mile. This meant that 2016 Olympic 800-meter silver medalist, Burundi's Francine Niyonsaba, was ineligible to compete in her signature event in Tokyo but was eligible for the 5,000-meter race. Namibia's Christine Mboma was excluded from the 400-meter competition but went on to win silver in the 200-meter event. Indian sprinter Dutee Chand, who successfully overturned an earlier disqualification due to her endogenous testosterone levels, was also in Tokyo to compete at 100 meters.

Because the IOC instructs individual International Federations to set their own standards, the qualifications to compete in women's sport vary from one sport to another. The qualifications also differ by the level and scope of competition. And they change over time, which indicates, as experts have long argued, that there is no definitive or universal way to regulate sex. Unlike weight categories, no standardized measuring device resolves sex classification.

To be clear, regulating sex is really about regulating femaleness. Few elite sport policies attempt to define and control who can compete in men's sports. And despite the important differences between transgender women and women with differences of sex development (DSD) or intersex variations, elite sport continues to draw them together with policies that define the criteria by which they can compete in women's events. This inevitably subjects sportswomen to added surveillance and scrutiny. As bioethicists Lance Wahlert and Autumn Fiester assert, "intersex athletes, transgender athletes, and *all* female athletes are intrinsically and woefully intertwined" by the regulation of sex in elite sport.[3]

Proponents of regulating femaleness maintain that it is imperative to "place conditions" on who can compete in order to "protect" women's sports and "ensure fair and meaningful competition." This is precisely what World Athletics (then the International Association of Athletics Federations) argued in 2019 when Semenya appealed to the Court of Arbitration for Sport to ask that the testosterone policy be declared "invalid and void with immediate effect."[4] The arbitrators ruled against her. They conceded that the regulations were "discriminatory," but that "such discrimination is a necessary, reasonable and proportionate means of achieving [World Athletics'] aim of preserving the integrity of female athletics."[5]

The concept of "necessary discrimination" in sport is an interesting one (so is "integrity," but that's a different kettle of fish). The Olympic Charter asserts that sport is a human right to which everyone should have access "without discrimination of any kind."[6] In that respect, discrimination based on gender diversity is anathema to the tenets of Olympism. Indeed, the United Nations Human Rights Council submits that World Athletics' regulations "may not be compatible with international human rights norms and standards." More urgently, Human Rights Watch insists that "global regulations that encourage discrimination, surveillance, and coerced medical intervention on women athletes result in physical and psychological injury and economic hardship" and "violate fundamental rights to privacy, health, and non-discrimination."[7]

Yet World Athletics holds that it "is not subject to human rights instruments" because it "is not a public authority, exercising state powers, but rather a private body exercising private (contractual) powers."[8] The federation invoked the "longstanding concept of 'sport exceptionalism,'" write Ali Durham Greey and Helen Jefferson Lenskyj, "that is, rules applying to other social contexts and workplaces must be suspended in relation to sport, so that women's 'safety' and 'fairness' may be guaranteed."[9] In international track and field, sporting rights supersede human rights.

Still, one might argue that necessary discrimination is foundational to sport. We discriminate on the basis of age, weight, impairment, and enhancement to create safe and fair competition. In the same way, we segregate elite sport by sex because, on average, elite male athletes outperform elite female athletes by an average of 10 to 12 percent.[10] Without discrimination *between* elite men and women's sports, at least in their current models, women would lose out. At issue here, how-ever, is discrimination *within* women's sport and who is allowed to compete in that category.

This chapter is riddled with tensions: between protecting one group and leaving another vulnerable; between necessary and harmful discrimination; between sporting rights and human rights; between "natural" bodies and the unnatural conditions of sport. As this brief history shows, policymakers have sought to resolve sexual classifica-tion with regulations concerning genitals, gonads, chromosomes, and genes. Each has been struck down as a definitive marker of sex. Will testosterone meet the same fate? Or are there viable alternatives to sport's two-sex system?

An Early History of Sex Testing

Modern organized sport developed during the nineteenth and early twentieth centuries as a distinctly "male preserve"—a space where men could cultivate, perform, and prove their masculinity amid the tumult of a changing society.[11] Sport purportedly taught boys and men the skills and character they needed to succeed in life. It helped them exercise and build their bodies at a time when the average workday was increasingly sedentary. Sport afforded opportunities for male camaraderie apart from "feminizing" influences and gave them the chance to test their mettle within a homosocial environment.[12]

Consequently, the qualities of what made one a "real athlete" became synonymous with what made one a "real man." Those same masculine-athletic qualities were rendered incompatible with turn-of-the-century womanhood. Women showing athletic interest or promise were regarded with suspicion, if not derision, and their participation signaled that they must not be "real women." Alternatively, sport would destroy "real womanhood" by virtue of its "masculinizing" or "virilizing" properties.[13] These foundations contribute to some of elite sport's deep problems.

Despite the proscriptions stacked against them, women began to forge their own organized sporting opportunities. They made their first Olympic appearance in 1900 when twenty-two women (2.2 percent of all Olympians) competed in the "gender-appropriate" pursuits of tennis and golf.[14] However, they were continually stonewalled in their efforts to compete in so-called masculine sports, especially track and field. Critics decried women's athletics as "profoundly unnatural" and charged that competitors would "develop wholly masculine physiques and behavior traits."[15] Others worried that "masculine" women naturally gravitated to the track. Indeed, anxieties about regulating sex have been most acute in track and field, which has routinely set the pace for other sports to follow.[16]

In the mid-1920s the IAAF consented, albeit grudgingly, to sponsor women's competitions. The IOC followed suit and allowed five women's track and field events at the 1928 Summer Olympic Games in Amsterdam. There, spectators viewed women who faltered on the track as proof they were not suited for the sport. By that same logic, women who succeeded must not really be women. This type of gossip dogged Japan's 800-meter silver medalist, Hitomi Kinue, who had been reportedly subjected to a two-hour examination "before it was decided the predominant sex."[17] The Japanese press fueled these rumors by estimating that Hitomi was "40 or 50 percent male and 50 or 60 percent female."[18]

American Olympic Committee president Avery Brundage likewise surmised that sex could be gauged by percentages. Shortly before the

1936 Summer Games in Berlin, he "roundly recommended that all women athletes entered in the Olympics be subjected to a thorough physical examination to make sure they were really 100% female." *Time* reported that Brundage was particularly vexed by "two athletes who recently competed in European track events as women were later transformed into men by sex operations."[19] Yet the athletes in question were not "transformed into men," but had rather been assigned female at birth, experienced differences of sex development, and underwent gender-affirming medical care.[20] Just the same, their sensationalized stories stoked misgivings about sportswomen.

The IOC Congress met just before the 1936 Games in Berlin to discuss what to do about "abnormal women athletes." Each International Federation was ultimately allowed to decide whether and how to test for sex, an option the IAAF evidently exercised on the American 100-meter champion Helen Stephens. After beating her rival, Poland's Stanisława Walasiewicz (who also went by her Americanized name Stella Walsh), Polish newspapers charged that Stephens was "really a man." The accusation, responded the American press, "brought the disclosure . . . that German Olympic officials had ascertained her true sex before admitting her participation in the 1936 games."[21]

The following year, the IAAF instituted what historian Sonja Erikainen concludes was "the first official sex testing policy instituted in elite sport."[22] The test was not comprehensive; that is, it was not applied to all women. Instead, the IAAF 1937 rulebook stipulated that if there were a "protest" concerning an athlete's "physical nature" a physician may conduct an inspection to verify her sex (see Table 2.1 for a timeline of sex-testing policies).[23]

The sporting pause during World War II did little to quell anxieties about female athletes. When international competitions resumed in 1946, the IAAF required anyone competing in women's events to present "medical certificates" confirming that they were, in fact, women. The IOC did the same for the 1948 Summer Games in London, although neither organization detailed the criteria for female certification.[24]

Table 2.1 Regulations and Decisions Regarding Sex Testing and the Participation of Transgender Athletes

1937	IAAF initiates first official sex testing policy
1946	IAAF requires a physician's letter to verify the sex of women competitors
1948	IOC requires a physician's letter to verify the sex of women competitors
1966	IAAF requires gynecological examinations at the British Empire and Commonwealth Games
1966	IAAF requires visual inspection of women athletes at the European Athletics Championships
1967	IAAF inaugurates sex chromatin test (Barr body) at the European Cup
1967	IAAF disqualifies Polish sprinter Ewa Kłobukowska based on the results of her sex chromatin test
1968	IOC trials the sex chromatin test at the Winter Olympic Games in Grenoble, France
1968	IOC implements the sex chromatin test at the Mexico City Summer Olympic Games
1976	Renee Richards, a transgender woman, applies to compete in the US Open; the United States Tennis Association and United States Open Committee respond by requiring the sex chromatin test for all women applicants
1977	Claiming a violation of the New York State Human Rights Law (Section 297(9) of the New York State Executive Law) and the 14th Amendment to the US Constitution, Renee Richards seeks a preliminary injunction against the United States Tennis Association, United States Open Committee (USOC), and the Women's Tennis Association to qualify for the 1977 US Open. The New York Supreme Court rules that the use of the test to disqualify transgender women is "grossly unfair, discriminatory and inequitable, and violative of plaintiff's rights," allowing Richards to play in the women's division of the tournament.
1985	María José Martínez Patiño is disqualified by the sex chromatin test at the World University Games; Martínez Patiño fights her disqualification
1988	IAAF reinstates Martínez Patiño's eligibility to compete in women's events
1990	IAAF-sponsored Workshop on Methods of Femininity Verification; Council recommends the replacement of laboratory-based tests with "a medical examination for the health and well-being of all athletes (women and men)"
1991	IAAF Council replaces "gender verification" with "a medical examination for the health and well-being of all athletes (women and men) that would "obviate the need for any laboratory-based genetic 'sex test'"
1992	IAAF Working Group recommends discontinuing medical examinations; IAAF approves recommendation

Table 2.1 Continued

1992	IOC uses polymerase chain reaction tests for the Y-linked SRY gene (sex-determining region Y) at the Winter Olympic Games, Albertville, France
1996	At the Summer Olympic Games in Atlanta, Georgia, eight women athletes initially "fail" the SRY test, but after further examination, all are allowed to compete
1999	IOC Executive Board announces the end of comprehensive sex testing
2004	IOC's Statement of the Stockholm Consensus on Sex Reassignment in Sports
2011	IAAF Regulations Governing Eligibility of Females with Hyperandrogenism to Compete in Women's Competition
2012	IOC Regulations on Female Hyperandrogenism; sets women's endogenous testosterone at 10 nmol/L
2012	Four women diagnosed as hyperandrogenic are referred to a specialist reference center in France and subsequently undergo gonadectomies and "feminizing" cosmetic procedures
2014	Athletics Federation of India disqualifies sprinter Dutee Chand on the grounds of atypical testosterone levels
2014	Chand appeals her disqualification with the Court of Arbitration for Sport
2015	*Interim Arbitral Award: Dutee Chand v. Athletics Federation of India and the International Association of Athletics Federations*; Court of Arbitration for Sport temporarily suspends hyperandrogenism policies
2015	IOC Consensus Meeting on Sex Reassignment and Hyperandrogenism
2017	Publication of Stéphane Bermon, Angelica Lindén Hirschberg, Jan Kowalski, and Emma Eklund, "Serum Androgen Levels Are Positively Correlated with Athletic Performance and Competition Results in Elite Female Athletes" *BJSM* 52, no. 23 (2018): 1531–1532. Despite significant research flaws, this serves as the foundation for the IAAF's 2018 regulations.
2018	IAAF's Eligibility Regulations for the Female Classification (Athletes with Differences of Sex Development) replace hyperandrogenism policy and dismiss Chand's case; women racing between 400m and 1 mile must exhibit: testosterone below 5 nmol/L; sensitivity to testosterone; XY chromosomes; specific DSD diagnoses
2018	Caster Semenya appeals to the Court of Arbitration for Sport arguing that the IAAF's DSD regulations are "unlawful" and violate "universally recognized human rights," and to "prevent them from being brought into force on the basis that they are unfairly discriminatory, arbitrary and disproportionate." The arbitrators uphold the DSD regulations.
2019	World Athletics Eligibility Regulations for Transgender Athletes

(continued)

Table 2.1 Continued

2020	Semenya appeals the 2018 CAS decision to the Swiss Federal Tribunal; the tribunal dismisses her case
2021	Semenya appeals to the European Court of Human Rights to end "discriminatory" testosterone limits imposed on women athletes
2021	The *British Journal of Sports Medicine* prints a correction for the 2017 study by Bermon et al.
2021	IOC Framework on Fairness, Inclusion, and Non-Discrimination on the Basis of Gender Identity and Sex Variations
2023	World Athletics updates its Eligibility Regulations for Female Classification (Athletes with Differences of Sex Development) lowering the accepted endogenous testosterone threshold to 2.5 nmol/L in all events; it sets the same threshold in the updated Eligibility Regulations for Transgender Athletes: "Since puberty they must have continuously maintained the concentration of testosterone in their serum below 2.5 nmol/L."
2023	The European Court of Human Rights finds that the Swiss state discriminated against Semenya and that there were "serious questions" about the World Athletes' policy; World Athletics' DSD policy remained in place

As the Cold War blasted its chill through the world of sport, Eastern bloc women emerged as formidable athletes. Defying white, Western gender norms, Soviet women and their comrades in satellite nations engaged in rigorous, scientific training that launched their nations to the top of the Olympic medal count. While this encouraged other nations to develop opportunities for girls and women, it also cast aspersions on athletes "from Communist countries," who were, reported the *New York Times*, "of questionable femininity."[25]

These aspersions manifested in four interrelated charges. First, gender panic persisted over athletes who "changed sex."[26] Second, there were unsubstantiated theories that men were "masquerading" as women.[27] Third, there were new anxieties about doping and the "virilizing" effects of anabolic steroids (see Chapter 5).[28] Fourth, there were allegations that national programs intentionally recruited women athletes of "questionable sex." IOC executive Monique Berlioux, for example, argued that sex testing was necessary to identify athletes with a "sexual mutation." This would "put an end to the cheating . . . Nothing is more prejudicial to female sport [than] this charlatanry; nothing can kill it more surely."[29]

Fortified by these concerns, the IAAF initiated the first comprehensive sex test at the 1966 British Empire and Commonwealth Games. There, women were made to undergo gynecological examinations, which pentathlete Mary Peters described as

> the most crude and degrading experience I have ever known in my life. I was ordered to lie on the couch and pull my knees up. The doctors then proceeded to undertake an examination which, in modern parlance, amounted to a grope. Presumably they were searching for hidden testes. They found none and I left. Like everyone else who had fled that detestable room I said nothing to anyone still waiting in the corridor and made my way, shaken, back to my room.[30]

Later that year at the European Athletics Championships, physicians performed visual scans. The "nude parade," *Time* reported, was "perfunctory. Lined up in single file, the 234 female athletes paraded past three female gynecologists. 'They let you walk by,' said one competitor afterward. 'Then they asked you to turn and face them, and that was it.'"[31] All women apparently passed the test, although five athletes opted not to attend the event, several of whom promptly retired from athletics. This only emboldened officials. The test, crowed IAAF president David Burghley, "has been successful in frightening the doubtful ones away."[32]

Even so, there was a clear need for a less invasive method to verify competitors' sex. Chromosomes seemed to be the answer. In the mid-1950s, scientists discovered that humans typically have 46 chromosomes, which can be sorted into homologous pairs. The first twenty-two pairs are the numbered autosomal chromosomes; the twenty-third pair is the sex chromosomes. The typical male karyotype (an individual's collection of chromosomes) is 46,XY; for females, it is typically 46,XX, although many variations can occur.

During this same time, microanatomist Murray Barr discovered the Barr body—the second inactive X in the typically female sex chromosome pair—and devised methods to test for its presence. By the mid-1960s, physicians began to use the Barr body test "in the medical interpretation and management of the sexually anomalous," writes scholar Fiona A. Miller.[33] Sport similarly seized on the test to establish what one physician called "a new definition of femaleness."[34]

IAAF officials inaugurated the sex chromatin test at the 1967 European Cup, where clinicians swabbed competitors' cheeks and analyzed the sample under a microscope. Those whose tests showed the Barr body earned their "certificate of femininity"—a license confirming their femaleness—that was "valid for any amateur or international competition in which the athlete may participate after her examination."[35] Anything other than XX spelled disqualification.

Such was the case for Polish sprinter Ewa Kłobukowska, who passed her visual inspection in 1966. The following year, analysts ruled that she had "one chromosome too many," later presumed to be 46, XY/47, XXY mosaicism, in which a person's cells are genetically different.[36] Although the condition probably had little bearing on her athletic performance, Kłobukowska could no longer compete in women's track and field. "It's a dirty and stupid thing to do to me," said the twenty-three-year-old world champion at the time. "I know what I am and how I feel."[37] Yet throughout the history of regulating sex, what an athlete knows and how she feels have been largely irrelevant. As one Olympic physician rationalized, "A lady can not be a lady and not know it."[38]

The IOC's nascent Medical Commission formally adopted the sex chromatin test in 1968. Dr. Eduardo Hay, the commission's chief of testing, explained that the exam "verifies that the athletes are competing on an equal basis considering their physical status. In the cases of intersexuality or hermaphroditism, the athlete must be barred from competition in order to insure [sic] fair play."[39]

Despite the IOC's confidence in the sex chromatin test, members of the international medical community openly criticized its use. It was "grossly unfair" and "should not be used as absolute criteria of sexual identity," stressed anatomist Keith L. Moore, who had studied under Murray Barr.[40] Barr himself would later urge sports officials to abandon the test. Its application to "testing in the area of athletics is totally inappropriate," he wrote to Olympic officials. "Its use in this way has been an embarrassment to me and I request that it be stopped."[41] Instead, the test remained in place for thirty years.

It is not known how many women the tests disqualified or dissuaded from sport. Following Kłobukowska's ousting, IOC regulations stipulated that the results of the examinations "will not be made public out of deference to the human rights of the individual."[42] It was only when a woman openly challenged her disqualifying test results that these stories came to light. Enter Spanish hurdler María José Martínez Patiño, who initially passed the Barr body test in 1983. Two years later, at the 1985 World University Games, Martínez-Patiño forgot her certificate and was retested, this time with different results. Officials counseled she drop out of the event while they reviewed her case. She complied. "They wrapped my ankle to pretend I was hurt," she recalled, "but inside I was destroyed."[43]

Two months after that, Martínez Patiño received a letter informing her that her "Karyotype is decided 46, XY."[44] The IAAF had classified her as male and declared her ineligible for women's sport. She was shocked. "I could hardly pretend to be a man," Martínez-Patiño later wrote. "I have breasts and a vagina. I never cheated. I fought my disqualification."[45]

During that fight, she learned that she has complete androgen insensitivity syndrome (CAIS). Androgens are sex hormones that include testosterone. Individuals with CAIS typically have XY chromosomes, internal testes, and produce testosterone levels in the male-typical range, but their cells do not form typical androgen receptors and cannot respond to that testosterone. As Martínez Patiño explains, "When I was conceived, my tissues never heard the hormonal messages to become male."[46] She was, as one expert put it, "disqualified for having an advantage that she didn't have."[47]

The IAAF eventually restored Martínez Patiño's right to compete as a woman in 1988, tacitly admitting that women can have
XY chromosomes, or any combination of sex chromosomes for that
matter, and that different bodies respond differently to testosterone.
But it all came too late and proved too costly for the young athlete
at what should have been the peak of her career. "I was expelled
from our athletes' residence, my sports scholarship was erased from
my country's athletics records. I felt ashamed and embarrassed. I lost
friends, my fiancé, hope and energy," Martínez Patiño recounts. "I
paid a high price. . . . My story was told, dissected, and discussed in
a very public way."[48] This is important. The history of sex testing is
often traced from one athlete's "scandal" to the next. In the process,
we cavalierly, even callously trade in the secrets of their most intimate
details.[49]

The IAAF dropped the sex chromatin test in 1988 as a result of
Martínez Patiño's appeal and pressure from the international medical community.[50] The IOC did the same in 1992 but put in its place
a test for the SRY gene, found on the Y chromosome and linked to
male sexual development.[51] At the 1996 Olympic Games in Atlanta,
eight women out of 3,387 (1 in 423) exhibited the SRY gene. Medical
authorities cleared all eight women to compete as women due to
intervening conditions.[52] With medical censures calling the genetic
test "morally destitute" and a "futile exercise causing embarrassment,
anguish and expense," the IOC announced the end of systematic sex
testing in 1999.[53]

Despite the problems associated with sex testing, many athletes
support its use. Mary Peters, who decried the gynecological exam
she endured in 1966, nonetheless found it "satisfying to know . . . that
when you've been beaten you've been beaten by a genuine woman.
That wasn't always the case in my career before the introduction of
sex tests."[54] In 1994, sixteen Olympic and international-class women
runners petitioned the IAAF to reintroduce testing at "high-stakes
events."[55] Two years later, at the Atlanta Summer Olympic Games,
82 percent of female athletes who responded to a survey expressed

their support for sex tests and there remains a significant number of athletes and women's sports advocates who lobby for the necessity of some type of "sex control."[56] Most recently, this control has turned to testosterone (T).

Much Ado About T

An end to systematic sex testing in the new millennium did not mean an end to all sex testing in sport. Rather, regulatory bodies reverted to where they started seven decades earlier: by reserving the right to inspect any athlete in the event of a "protest" concerning her "physical nature." As articulated in the IAAF's 2006 Policy on Gender Verification, if there was "any 'suspicion,'" "or if there is a 'challenge' then the athlete concerned can be asked to attend a medical evaluation."[57]

There are justified criticisms of which athletes are most often deemed suspicious—about what women seem "masculine" when judged by white, Western gender norms. Specifically, since the end of comprehensive testing, there have been a disproportionate number of "challenges" levied against women of color from the Global South, as intimated in the list of athletes noted earlier as "conspicuously absent" from the 2020 Tokyo Games.[58]

Santhi Soundarajan was among those challenged. After she finished second in the 800-meter race at the 2006 Asian Games, the Indian Olympic Association announced that she "did not possess the sexual characteristics of a woman." Without explaining what those characteristics were, officials stripped her of her silver medal and banned her from future competitions for a "Games rule violation."[59] Soundarajan later attempted suicide, explaining to reporters, "I am physically and mentally totally broken."[60]

Caster Semenya likewise fell afoul of suspicion-based testing at the 2009 World Championships. As the eighteen-year-old cruised to victory in the 800-meter race, IAAF officials confirmed that she had been subjected to "gender verification" procedures.[61] Secretary Pierre Wiess recalled a 1936 Avery Brundage when he confided to reporters that Semenya "is a woman but maybe not 100 per cent."[62]

The federation allowed Semenya to retain her title and prize money but requested she withdraw from competition while a panel of experts reviewed her tests.[63] After nearly a year, the IAAF announced that she could once again "compete with immediate effect."[64] Semenya was noticeably slower upon returning to the track, which fed suspicions that she had undergone some type of treatment to reduce her testosterone levels.

Months later, the 2011 Regulations Governing Eligibility of Females with Hyperandrogenism to Compete in Women's Competition all but confirmed those suspicions. The IAAF announced that it had "abandoned all reference to the terminology 'gender verification' and 'gender policy' in its Rules."[65] The IOC issued nearly identical guidelines in advance of the 2012 Summer Games in London, insisting that "nothing in these Regulations is intended to make any determination of sex."[66] These are just semantics. Setting biological limits on who can compete in women's sports and designing methods to assess that biology is, indisputably, a determination of sex.

The regulations revolved around women's "hyperandrogenism," defined in the IAAF's 2011 policy as "the excessive production of androgenic hormones (Testosterone)," identified as functional endogenous testosterone above 10 nmol/L.[67] In healthy adult cismen, typical testosterone levels range from 7.7 to 29.4 nmol/L, and in healthy adult ciswomen from 0.12 to 1.79 nmol/L. While experts disagree on the specific cut points and acknowledge a significant "overlap of endogenous testosterone levels between the sexes," the biggest point of contention has been the association of testosterone with athletic performance.[68]

On one side, a renowned group of scholars asserts that *"there is no evidence showing that successful athletes have higher testosterone levels than less successful athletes."*[69] On the other side, an equally renowned group concedes the point: "There are few published studies on the influence of endogenous testosterone and athletic performance in women." They nevertheless maintain that the "male advantage in certain sports is most likely explained by the fact that men produce much higher levels of androgenic hormones, most notably testosterone."[70] Is "most likely" good enough to enact policy?[71] It depends on who you ask.

The IOC and IAAF apparently thought that hyperandrogenism was not only performance-enhancing but also posed a health risk to athletes. They described it as a pathological "disorder" in need of "prescribed medical treatment," presumably surgical and/or pharmaceutical intervention that would lower women's testosterone.[72] Yet there is no evidence that hyperandrogenism, on its own, poses a risk to athletes' health.[73] Actually, the "treatment" may do more harm than the "disorder" itself.

This possibility reared its ugly head when it was revealed that IAAF-affiliated physicians diagnosed hyperandrogenism in four athletes, aged eighteen to twenty-one, from "rural and mountainous regions of developing countries." The women were sent to a "specialist reference center" in France, where physicians recommended they undergo gonadectomies (the removal of testes). In a subsequent publication, the physicians wrote that although the presence of "male gonads . . . *carries no health risk*, each athlete was informed that gonadectomy would most likely decrease their performance level but allow them to continue elite sport in the female category." Doctors also performed a series of cosmetic procedures on the women that had no bearing on their health or sports eligibility, including "feminizing vaginoplasty and estrogen replacement therapy, to which the 4 athletes agreed after informed consent on surgical and medical procedures."[74]

In a scathing rebuke, a multidisciplinary research team disputed the idea of "informed consent." "Given that [the women's] eligibility to compete was clearly dependent upon agreeing to the procedures, the line between consent and coercion is blurred in this instance." They further stressed that the "medical decisions rendered violate ethical standards of clinical practice and constitute a biomedical violence against their persons."[75] Ugandan runner Annet Negesa agreed. As she wrote years later, "I believe I was one of the four athletes referred to in that article."

Negesa has since described the "intrusive" tests she took in France and her referral to a surgeon. "I was told that if I wanted to keep running, I needed to have a medical intervention. At the time, I thought it was going to be like a needle pulling fluid from my body." Instead, she came out of anesthesia "with bandages on [her] abdomen" and learned that she would "need life-long hormone therapy" after a gonadectomy to which she says she did not consent. As she writes, "My body was shamed, humiliated, surgically altered even though I didn't need it to be, and used as an object of study simply because I am different and I wanted to run."[76]

It wasn't until 2019, seven years after the ordeal, that Negesa met activist Payoshni Mitra and "learnt, in detail, about the regulations that had pushed me out of my sport." Dr. Mitra began her tireless advocacy in 2009 with Santhi Soundarajan and, over time, has helped support many athletes. This includes Dutee Chand, who the Athletics Federation of India disqualified from competition in 2014. Chand refused to either drop out of sport or submit to medical intervention. Instead, with the support of Mitra and others, she appealed her disqualification in the Court of Arbitration of Sport (CAS). "The high androgen level produced by my body is natural. I have not doped or cheated," she testified. Experts advised her that medical intervention to lower her testosterone was "unscientific," "invasive," "often irreversible and will harm my health now and into the future."[77]

The IAAF called a slate of witnesses to refute Chand's claims. "Differences in testosterone are the most significant factor in explaining the performance differences between male and female athletes," the legal team argued, citing the 10 to 12 percent performance gap between elite male and female athletes in comparable events.[78] As such, it is imperative to segregate the sexes by testosterone "for reasons of fairness."

The CAS arbiters agreed with the IAAF's premise, but found that the IAAF had "insufficient evidence about the *degree* of the advantage that androgen-sensitive hyperandrogenic females enjoy over non-hyperandrogenic females." Specifically, the IAAF lacked the scientific backing to argue that hyperandrogenic women enjoyed "commensurate significance to the competitive advantage that male athletes enjoy over female athletes."[79] Women with high endogenous T were not turning in "male" performances. The CAS's "Interim Award" temporarily suspended the regulations and gave the IAAF two years to establish such evidence.

To the consternation of the IAAF and its supporters, "hyperandrogenic" women could compete at the 2016 Summer Olympic Games in London and other events without suppressing their endogenous testosterone. Joanna Harper, who describes herself as "a scientist first, an athlete second, and as a transgender person thirdly," was among those who voiced their disapproval. "While human rights advocates are deliriously happy over the CAS ruling, those who love women's sport are mortified," she decried. "Those Intersex athletes who previously used medications to reduce their T are now off of those medications, and are running faster. Allowing these athletes to compete in women's sport . . . threatens the very fabric of women's sport."[80]

Behind the scenes, the IAAF scrambled to protect their proprietary pattern in that fabric. But rather than returning to the CAS with evidence to support the hyperandrogenism regulations, the IAAF withdrew them entirely and devised a new policy to take effect in 2018. This effectively dismissed Chand's case because, as a sprinter, she was not considered a "relevant athlete." Instead, the 2018 Eligibility Regulations for the Female Classification (Athletes with Differences of Sex Development) only applied to women who competed in "restricted events"—international races between 400 meters and 1 mile. Only middle-distance runners would now have to show serum testosterone equal to or less than 5 nmol/L, half of what the previous hyperandrogenism policy allowed. In addition, they must be sensitive to testosterone (not have complete androgen insensitivity syndrome), exhibit XY chromosomes, show the presence of testes or testicular tissue, and be diagnosed with a specific DSD.[81] High endogenous testosterone is not the only biomarker taken into account, but it is the first step in identifying possible "relevancy." It is also the only condition that must be changed for an athlete to compete as a woman.

The foundation for the regulations was a 2017 study that was funded by the IAAF and conducted by IAAF-affiliated researchers.[82] Unsurprisingly, the study's ethics, design, methodology, analysis, and conclusions have been roundly criticized.[83] In fact, shortly after the 2020 Tokyo Games, the *British Journal of Sports Medicine* published a correction to that study, in which the authors acknowledged that it was "exploratory" and "that statements in the paper could have been misleading. . . . To be explicit, there is no confirmatory evidence for causality in the observed relationships reported."[84] In lay terms: the evidence did not verify that higher testosterone causes better performance. The original statements were more than "misleading"; they were scientific malfeasance. Still, they remain the bedrock of the federation's policies on eligibility for women's sport.

Semenya refused to alter her natural body, a curious request, considering that one of the rationales against doping is that it is "unnatural." She had suppressed her testosterone between 2010 and 2015, she explained, and the drugs made her "sick constantly." She experienced weight gain, fevers, abdominal pain, and "enormous" mental anguish. There were also twice-monthly blood tests to monitor her testosterone, all of which made her feel like a "lab rat."[85]

Turning to her only recourse, the CAS, Semenya argued that the IAAF discriminated against women "based solely on a natural or genetic trait which they have possessed since birth and over which they have no control." She further submitted that the "world celebrates the genetic differences that make athletes such as Usain Bolt, Michael Phelps and Serena Williams great. DSD are a form of genetic difference that should be celebrated in the same way."[86]

The IAAF countered that sport is not segregated by the kinds of genetic differences enjoyed by the likes of Bolt, Phelps, and Williams, but it *is* segregated by sex. The crux of their defense was that the "relevant athletes" were not "biologically female," but rather "biologically male athletes with female gender identities" and that "there are some contexts where biology has to trump identity."[87] Yet it is worth noting that the "biological males" in question were not turning in "male performances" that were 10 to 12 percent better than "biological females." At her best, Semenya beat competitors by about 2 percent. Sport scientist Ross Tucker calculated her average margin of victory at 1.03 percent.[88] While 1 to 2 percent will distinguish a champion from an also-ran, her times do not come anywhere near elite men's times.[89]

After two months of deliberation, a three-person panel voted 2–1 in agreement with the IAAF: discrimination was "necessary, reasonable and proportionate" for "the upholding of the 'protected class' of female athletes in certain events."[90]

Payoshni Mitra was there. Although "devastated" by the ruling, she emphasizes that Semenya, Chand, Negesa, and others are "heroic." As Mitra explains, there tends to be "this Western lens where we constantly try and look at them as victims. I don't look at them as victims. They are winners." Too often lost in this history, she stresses, is "the way these athletes have fought back. . . . This is a story of hope and resilience. It's not a story of victimhood. There is violation, of course, but the fight back is bigger."[91]

Semenya's fight back cost her the chance to defend her 800-meter Olympic title in Tokyo 2020, but the outcome of that race provides an interesting point of comparison (Table 2.2). All three medalists in 2016, including Semenya, have since disclosed that they were "relevant athletes" under the DSD regulations; however, due to the CAS's 2015 "Interim Award," they were able to compete in Rio without suppressing their endogenous testosterone. Each one of their times is *slower* than their 2020 medal-winning counterparts, who all ran in accordance with the IAAF's DSD regulations. In short, women with endogenous testosterone below the 5 nmol/L threshold were faster than those presumed to be above it.

This is not to suggest that testosterone is meaningless to athletic performance.[92] The problem is that we don't know the degree of

Table 2.2 Women's and Men's 800-Meter Medalists, 2016 and 2020 Summer
Olympic Games

	Women's 800m Winners, 2016 Rio Summer Games	Women's 800m Winners, 2020 Tokyo Summer Games	Men's 800m Winners, 2016 Rio Summer Games	Men's 800m Winners, 2020 Tokyo Summer Games
Gold	Caster Semenya (ZA), 1:55.28	Anthing Mu (USA), 1:55.21	David Rudisha (KEN), 1:42.15	Emmanuel Korir (KEN), 1:45.06
Silver	Francine Niyonsaba (BUR), 1:56.49	Keely Hodgkinson (GBR), 1:55.88	Taoufik Makhloufi (ALG), 1:42.61	Ferguson Rotich (KEN), 1:45.23
Bronze	Margaret Wambui (KEN), 1:56.89	Raevyn Rogers (USA), 1:56.81	Clayton Murphy (USA), 1:42.93	Patryk Dobek (POL), 1:45.39

Note: While the women's times at the 2020 Games were faster than in 2016, the men's times were slower, suggesting that the environmental conditions in Tokyo did not contribute to the women's improvement.

meaning that testosterone has. We may never know. Endocrinologist David Handlesman, who testified for the IAAF in Semenya's appeal, confided to a colleague that the "lack of substantial direct evidence" regarding high T on "female athletic performance will always be a severe limitation so we must rely on the best relevant surrogate evidence."[93] The surrogate evidence is derived primarily from medical studies that either introduce exogenous (foreign or artificial) testosterone in patients or suppress their endogenous testosterone.[94] Researchers cannot be certain that exogenous and endogenous T have the same effects or that those studies translate to elite athletics.

Undeterred, World Athletics again narrowed the scope of who can compete in women's events in the 2023 version of its Eligibility Regulations for Female Classification (Athletes with Differences of Sex Development). The regulations "require any relevant athletes to reduce their testosterone levels below a limit of 2.5 nmol/L for a minimum of twenty-four months to compete internationally in the female category in any event, not just the events that were restricted (400 meters to 1 mile) under the previous regulations."[95] The

2.5 nmol/L threshold also holds in World Athletics' 2023 Eligibility Regulations for Transgender Athletes.

Trans Policies

Trans-specific sport policies have a much shorter history than sex testing, although its various iterations likely disqualified transgender women over the years. It was for this reason that the United States Tennis Association introduced sex testing in advance of the 1976 US Open, when Renee Richards, a trans woman, announced her intent to play.

Richards refused the test and was denied entry. The following year, she sought a preliminary injunction against the rule and the New York Supreme Court ruled in her favor. The judge found that Richards was legally recognized as a woman and that the sex test was "grossly unfair, discriminatory and inequitable, and violative of plaintiff's rights." Discriminating against her wasn't "necessary." Richards went on to play in the 1977 US Open, where she lost in the first round of the singles tournament, reached the finals of the women's double tournament, and continued to play professionally until 1981.[96]

It was at a 1990 Workshop on Methods of Femininity Verification that the IAAF "first considered," transgender inclusion, although it "was discussed only in passing," recalled one insider.[97] There, experts recommended that trans women who had "undergone prepubertal sex reassignment would be allowed to compete in women-only events." For those who transitioned after puberty, the group advised "that the relevant sports authority assess individuals on a case-by-case basis."[98] The IAAF implemented this recommendation.[99]

The individualized approach changed in 2003, when the IOC Medical Commission organized an ad-hoc committee in Stockholm "to discuss and issue recommendations on the participation of individuals who have undergone sex reassignment (male to female and converse) in sport." Adopted by the IOC's Executive Board in 2004, the Statement of the Stockholm Consensus on Sex Reassignment in Sports endorsed the same three principles for both trans women and men:

- Surgical anatomical changes have been completed, including external genitalia changes and gonadectomy
- Legal recognition of their assigned sex has been conferred by the appropriate official authorities
- Hormonal therapy appropriate for the assigned sex has been administered in a verifiable manner and for a sufficient length of time to minimize gender-related advantages in sport competitions[100]

This was not policy but offered guidance to International Federations as they devised their regulations.

The Stockholm Consensus was "important," write sociologists Sheila L. Cavanaugh and Heather Sykes, because until then, "most sport governing bodies either had no policy designed to admit [trans] or intersexed persons into competition or defaulted into a reactionary 'female at birth' policy.'" At the same time, the consensus statement offered a "very narrow definition" of what it meant to be transgender that excluded "a large segment of the international [trans] community."[101] In particular, legal recognition and surgical intervention were neither available to nor desirable to all trans athletes.

With these critiques in mind, the IOC replaced the Stockholm Consensus with its 2015 Consensus Meeting on Sex Reassignment and Hyperandrogenism, bringing together trans and intersex athletes under the same umbrella. It dropped the preconditions of legal recognition and gender-affirming surgery on the grounds that it is "not necessary for fair competition and may be inconsistent with developing legislation and notions of human rights." Instead, it recommended the IAAF's testosterone threshold (10 n mol/L) for hyperandrogenic women as the standard for trans sportswomen. Conversely, athletes "who transition from female to male" could "compete in the male category without restriction."[102]

Over time, regulating sex has primarily come down to regulating T—at least for women. There is no limit to endogenous testosterone for male athletes. WADA allows cismen diagnosed with "androgen deficiencies" (hypogonadism) to apply for therapeutic use exemptions (TUEs). They can take an otherwise banned substance—exogenous testosterone—because they have a medical need. That same policy reads that a "TUE for androgen deficiency should *not be approved for females*."[103] Even if a cis woman's functional testosterone falls below a healthy limit, she is not granted the same opportunity as a man to raise her hormone levels commensurate with her same-sex competitors.

The same position holds for trans women, who are likewise prohibited from "testosterone supplementation," even when low T poses health concerns.[104] Simultaneously, WADA recognizes that testosterone therapy for trans men is "essential for the anatomical and psychological transition process" and that supplementation will be "life-long in transgender male athletes." Under WADA's TUE guidelines, trans men should "follow the general principle of hormone replacement treatment of [cis-] male hypogonadism."[105] Taken together, these policies link testosterone with both maleness and the regulation of sex.

Toward New and Different Frameworks

Since the 1940s, the influence of the IAAF/World Athletics has extended well beyond the track, as other federations have adopted its policies as their own. The IOC, too, has often fallen into regulatory lockstep with athletics, although its 2004 and 2015 consensus statements diverged slightly.

Those divergences widened dramatically in mid-November 2021 when the IOC announced its new Framework for Fairness, Inclusion, and Non-Discrimination on the Basis of Gender Identity and Sex Variations. The title alone suggests a sea change in the IOC's approach, which it formulated after consulting for over two years with more than 250 athletes and stakeholders across multiple identities, fields, sports, and countries. The document opens with the assertion that "every person has the right to practice sport without discrimination and in a way that respects their health, safety, and dignity."

Per that assertion, the IOC recommends a "principled approach" to policy development based on "a coherent whole" of ten key considerations:

1. Inclusion
2. Prevention of harm
3. Non-discrimination
4. Fairness
5. No presumption of advantage
6. Evidence-based approaches to regulation
7. The primacy of health and bodily autonomy
8. A stakeholder-centered approach to rule development
9. The right to privacy
10. Periodic review of eligibility regulations[106]

This is, in the end, a human-rights approach to guiding policy.

Fittingly, the new framework confronts several disputed protocols. For example, it counsels organizations against "invasive physical examinations" and discourages "medically unnecessary procedures of treatment" for athletic eligibility. The framework also recommends that "until evidence (per principle 6) determines otherwise, athletes should not be deemed to have an unfair or disproportionate competitive advantage due to their sex variations, physical appearance and/ or transgender status." Put differently, that evidence does not yet exist, and until it does, presumptions of advantage are not enough to dictate policy.

"I feel that this is a huge achievement," reflects Dr. Mitra, who was among those the IOC consulted. "For the first time it feels like the IOC is speaking the same language that we have been speaking for so long." Altogether, supporters hope that the ten principles will "herald a new era of gender-inclusive sports participation and governance."[107]

Not so fast, said World Athletics, which affirmed that its "DSD and transgender regulations remain in place" just hours after the IOC announced its framework. Because the IOC's document is not "legally binding," one official explained, International Federations may continue to devise their own policies.[108] It therefore remains to be seen whether organizations will work within the IOC's framework, adopt World Athletics' regulations, or take a different tack altogether.

Pundits have theorized what those tacks might look like. Some proposals keep the male-female categories intact but include "third gender" classes for athletes who do not align—or do not want to align—with the binary options.[109] A similar prospect would eliminate "coercive" sex segregation in favor of "voluntary" separation and allow individual athletes to choose in which category to compete.[110] Another suggestion involves a "handicap system" for trans and intersex athletes that factors in testosterone. They would be eligible for women's sports, but their performances would be penalized for high T, similar to handicaps in horse racing (Chapter 1). "By using a handicap system," writes bioethicist Andria Bianchi, "the appropriate winner can be determined while at the same time allowing trans women to fairly compete in female categories.[111] Yet fairness implies that athletes are competing under the same conditions, for which this alternative does not allow.

Conversely, sport psychologists Vikki Krane and Heather Barber argue that a "radical transformation of sport is necessary to become truly inclusive."[112] This would reimagine sport not by sex segregation, but by other factors deemed relevant to athletic performance.[113] One research team suggests a sport-specific "algorithm" that would account for both "social parameters including gender identity and socioeconomic status" and "physical parameters" that might include testosterone levels in addition to height, weight, hemoglobin levels, bone strength or structure, lung capacity, and heart size.[114]

There are related proposals that resemble the classification system used in Para sport (Chapter 3). Briefly, athletes within the same sport or event compete in different "sport classes," determined by how their impairments affect their sporting abilities. Adapting this approach to "unisex" sport might consider "10 categories of advantages" that "provide a competitive edge," such as lean body mass, height, muscle strength, and vision, suggests legal scholar Maayan Sudai. "We could then assign each athlete a numerical grade in relation to the sport they wish to compete in."[115] Those grades determine one's sports class.

These are interesting thought experiments, but it's hard to conceptualize their implementation. Athletic performance is a wonderful, wonderous amalgam of historical, cultural, social, political, biological, genetic, biomechanical, and physiological circumstances and any classification is bound to have its limitations. The bigger question is whether we want elite sport to be truly inclusive, or if "necessary discrimination" is the name of the game.

3

Regulating Impairment

Shortly after Ian Silverman was born, doctors diagnosed him with cerebral palsy, a form of brain injury or atypical development that can affect one's bodily movement, muscle control and coordination, posture, and balance. For Silverman, the condition meant multiple surgeries, full leg casts, various treatments, and physical therapy that, when he turned seven years old, brought him to swimming. He turned out to be good. Very good. Before long, he began training at the North Baltimore Aquatic Club, which has produced champions the likes of Allison Schmitt, Michael Phelps, and, as of 2012, Ian Silverman.[1]

Silverman won gold in the 400-meter freestyle at the 2012 Paralympic Games in London, competing in the S10 category. Two years later, at the Pan Pacific Para-Swimming Championships, Silverman broke the S10 freestyle world record. He continued to set records and collect victories until April 2015, when World Para Swimming, the sport's International Federation, told him he was no longer impaired enough to compete. His cerebral palsy hadn't changed. His classification had.

In Para swimming, as in many Para sports, competitors are grouped together according to their "degree of activity limitation," as World Para Swimming explains. This is to "ensure competition is fair and equal."[2] In Silverman's freestyle event, there are ten categories, S1 to S10. The S stands for swimming; the number is the "sport class" and the "lower the number, the more severe the activity limitation."[3] Athletes in the S1 class have the most limitation with "severely restricted movement" in all four limbs due to conditions such as quadriplegia or polio.[4]

Swimmers who compete in the S10 class have the least severe physical limitations, although not all have cerebral palsy. When Silverman set the S10 400-meter world record, he edged out Canada's Alec Elliot, diagnosed with syndactyly, a condition that limits his use of his hands and feet, and USA's Dalton Herendeen, whose left leg was amputated below the knee. Other S10 competitors might have severe restriction in one hip joint, double-foot amputation, or single-hand amputation.[5] They all compete in the same sport class because their conditions similarly affect their swimming abilities. Additional classes, S11 to S14, exist for athletes with visual and intellectual impairments.[6]

Then came the 2015 news that Silverman had been "classed out" of competition. "It hurts me a lot," he said at the time. "I don't understand it. Nobody understands it." His competitors agreed. "To me, this whole story is nonsense," said Benoit Huot, a nine-time Canadian Paralympic gold medalist (also in the S10 category because of club feet). Silverman "has a disability," Huot asserted. "What's the difference with Ian between now and 2012?"[7]

The difference, it seems, came down to the classifiers—the team of experts that decides in which class athletes compete—or whether they can compete in Para sport at all. Most likely, Silverman was a victim of classifier error. As he later wrote to the International Paralympic Committee (IPC), "the classifiers got it wrong . . . their assessments were contrary to my physical exam, x-rays, and reams of medical records."[8] He was right. Just over a year later, his diagnosis unchanged, Silverman was again "classed in" to Para swimming.

There is a lot to Silverman's story, not least of which is the social construction of "impairment"—the term the IPC and the World Health Organization prefer to "disability."[9] The meaning of the term may seem self-evident, but the role of classifiers shows that they have the power to determine who is impaired and who is not, at least for the purpose of sport. They enforce elite sport's protective policies and Silverman's story shows the fallibility of the process. Classifiers go through rigorous training, are expertly qualified and accredited, base their decisions on the best available evidence, and consult with others to make sure they get it right. Still, mistakes happen, and they remind us of the power that bureaucratic decisions have over athletes' careers.

Then there's the deceptive practice of "intentional misrepresentation." In fact, the main thrust of Silverman's letter to the IPC was about what he called "a deep-rooted problem with athletes misrepresenting themselves during classification to gain a competitive advantage in a lower class."[10] They pretend to be more impaired than they are so they have a better chance at winning. They cheat.

In response to intentional misrepresentation and other complications, World Para Swimming revised its classification rules in 2018. The new protocol puts more emphasis on observing the athletes in the water to assess their "swimming behavior." Said IPC Medical and Scientific Director Peter Van de Vliet, "Para swimming, as Para sport in general, is evolving all the time, and so is the knowledge of the technical aspects of the sport."[11] The process has corrected several problems and created better parity among swimmers, but complications remain.

Irish swimmer Ailbhe Kelly retired after the new rules, frustrated by the disparity between competitors in her S8 class. Even clocking her personal best in the 400-meter freestyle, she still finished thirty seconds behind the winner. She later told reporters, "it gets to the point where it's difficult to devote yourself to a sport when you know you can't compete in it properly."[12] Not long after that, Britain's Amy Marren, a world champion and Paralympic gold medalist, also quit competitive swimming. As she posted on Facebook, "there is a long way to go before it becomes a level playing field and this inconsistency across classifications is one reason why I am choosing to step away from a sport I have loved so very much."[13] The new regulations are, on the whole, better, but they still disadvantage and discriminate against some athletes—as any regulations do.[14]

Classification in Para sport is "unique," notes Van de Vliet. "It's the only thing that makes us different from able-bodied sport."[15] Indeed, the "para" in Para sport signifies that it runs "parallel" to so-called able-bodied sport—another problematic term. Just watch a snippet of the Paralympic Games and tell me those athletes' bodies aren't able.

Each Para sport (sometimes called disability sport or adapted sport) has its own classification system, set by the sport's International Federation. Sport classes are a relatively new development, spurred by the growing number of elite Para athletes and the need for defensible and transparent rationales for grouping them together in competition. According to the IPC, the "aim" of classification is "to ensure that the impact of Impairment is minimized and sporting excellence determines which Athlete or team is ultimately victorious."[16]

Researchers affirm that "Para sport classification systems provide a unique framework that permits Para athletes to demonstrate that elite athletic performance is a relative, rather than an absolute, concept."[17] *Citius, altius, fortius* is defined within the context of one's sport class. An S1 swimmer's performance should not be judged against the performances of swimmers in the S10 class, just as, in theory, the junior athlete should not be judged against the senior, the featherweight boxer should not be judged against the heavyweight, and women's performances should not be judged against men. In theory.

To be clear, and despite its complexities, classification is good for Para sport. "Without it," argues P. David Howe, a Paralympian and medical anthropologist, "Paralympic sport could not exist."[18] Classification has legitimized, equalized, and streamlined competition, which has brought in new spectators, media attention, and sponsorship. This is important not just for elite athletes, but for everyone. As Para sport grows in participation and popularity, it stimulates the development of more programs, promotes social interaction and inclusion, and breaks down the social stigma of impairment.[19] Yet as with the other protective policies explored in this book, sports classes also give rise to a host of unintended consequences, including those related to discrimination, autonomy, individual rights, and deceptive practices.

Concerns about classification are signs of growth and privilege, but they also mask deeper problems. The World Health Organization estimates that 1 billion people—15 percent of the world's population—live with some type of "disability," an admittedly "complex, dynamic, multidimensional, and contested" term.[20] Of those 1 billion people, 80 percent live in developing countries with limited social, financial, and health resources. War, political turmoil, disease, poverty, cultural stigma, race, gender, and social status, all intersect and compound issues related to impairment. According to the United Nations, only 45 of the world's 195 countries have anti-discrimination and other disability-specific laws to protect civic and human rights.[21] As scholar Simon Darcy points out, "in many parts of the world sport cannot even be contemplated because of the deplorable levels of poverty, unemployment and lack of human rights for people with disability."[22]

Consequently, a growing "disability divide" between developing and developed nations has left "a small group of behemoths" to dominate the global Para sport landscape.[23] A survey of the competitors at the 2015 IPC Athletics World Championships found that over half the participating nations were "High-Income Countries," based on the World Bank's classification; 44 percent were "Middle-Income Countries"; and just 4 percent were "Low-Income Countries."[24] These inequities bore out at the 2020 Paralympic Summer Games, which hosted 162 National Olympic Committees. The majority of those teams consisted of just two athletes, but the top four medal-winning federations—China, Great Britain, Russia, and the United States—each brought more than 220 athletes.[25] Together, these middle- and high-income nations swept a total of 553 (34 percent) of the total possible 1,617 at the Games.[26] Then again, a similar economic divide plays out in most elite sport, so maybe classification really is the only thing that makes Para sport "different from able-bodied sport."

A Brief History of Para Sport

Para sport has not always been organized according to sport classes, but rather by impairment type or medical diagnosis. The Berlin Sports Club for the Deaf, founded in 1888, was the earliest recorded organization of its kind. Later, the Comité International des Sports des Sourds/International Committee for Deaf Sports, established in 1922, was the first such international society; two years later it hosted the Games for the Deaf—now called the Deaflympics and staged every four years. In 1924, he Disabled Drivers Motor Club became the first athletic association for people with physical impairments. Most chronologies identify the second as the British Society of One-Armed Golfers, founded in 1932.[27]

World War II was a catalyst for much of the Para sport we know today. As the survival rate for people with spinal cord injuries vastly improved at mid-century, sport became a form of remedial treatment and rehabilitation that grew increasingly competitive and institution-alized. Since then, a range of sport- and impairment-specific associations have emerged to provide more competitive opportunities. The Special Olympics, for example, is the world's largest sports organization for children and adults with intellectual impairments.[28] The Athletes with Disabilities Network administers the Extremity Games, an extreme sport competition for athletes with limb loss or limb difference. There are also events for injured military personnel and veterans, including the Warrior and Invictus Games, as well as multiple sport- and impairment-specific events held at the local, regional, national, and international levels. All of these can be called "Para sport," although this chapter centers primarily on Paralympic sports and athletes with physical impairments.

The father of the Paralympic movement, Ludwig Guttmann, was a Jewish neurosurgeon who left Nazi Germany for England in 1939 (Figure 3.1). Four years later, he became the first director of Spinal Cord Injuries at the Ministry of Pensions Hospital, Stoke Mandeville,

Figure 3.1 Sir Ludwig Guttmann, unknown date, unknown photographer. Wikimedia Commons.

Aylesbury.[29] At the time, most of Guttmann's patients were servicemen and civilians whose spinal cord injuries caused paraplegia—partial or complete paralysis of the lower body.

Historically, there was a high mortality rate for people with serious spinal cord injuries. Approximately 80 percent of British and American soldiers who sustained them during World War I did not survive.[30] Yet Guttmann wrote that he saw his patients as more than "merely an accumulation of doomed individuals."[31] Rehabilitation, including physiotherapy, occupational therapy, and vocational training played a significant role in reintegrating the patients back into society to lead healthy, productive lives.[32]

Despite Guttmann's innovative approach to movement therapy, it was his patients who inspired his turn to sport. In 1944, the doctor came across a group of men playing a makeshift game of "wheelchair polo" with a wooden puck and walking stick. Polo, Guttmann concluded, was too dangerous, but he noticed the competition encouraged positive therapeutic aspects.[33] He subsequently promoted the gentler wheelchair sports of archery, netball, and javelin. As he would later write in his 1976 *Textbook of Sport for the Disabled*, sport was "the most natural form of remedial exercise, restoring physical fitness, strength, coordination, speed, endurance, and overcoming fatigue." It also "had a psychological impact of restoring pleasure in life and contributing to social reintegration."[34]

The 1940 and 1944 Olympics were cancelled during World War II, thus amplifying the significance of the 1948 Games. As the opening ceremonies began in war-torn London, fourteen men and two women representing Stoke Mandeville Hospital and the Star and Garter Home in Richmond, Surry, met in Aylesbury for the inaugural Stoke Mandeville Games for Paraplegics. Archery was the only sport on the program. The following year, Stoke Mandeville hosted thirty-seven athletes from six British hospitals who competed in archery and wheelchair netball. There, Guttmann delivered a speech in which he expressed hope that the games would become international and achieve "world fame as the disabled men and women's equivalent of the Olympic Games."[35]

Guttmann was prescient on both points. A team from the Netherlands joined the competition in 1952 for the first step toward internationalization. Over 130 athletes representing their countries, rather than their hospitals, competed in archery, darts, snooker, and table tennis. In 1960, the Games took place in Rome after the Summer Olympics, marking the first time the event was staged outside hospital grounds. In an event unofficially referred to as the "first Paralympics," 400 wheelchair athletes from twenty-three countries took part in eight sports: archery, athletics, basketball, fencing, snooker, swimming, table tennis, and dartchery (a combination of darts and archery).[36]

The Development of Sport Classes

The inclusion of athletes with different impairment types into the Paralympic Games initiated a "slow transition" from medically based classification, to sport-specific functional classification, and finally to the "model of best practice" used today.[37] The 1976 Paralympic Summer and Winter Games introduced sports for athletes with amputations and visual impairments. Four years later in Arnhem, the Netherlands, athletes with cerebral palsy competed in the 1980 Games. In 1984, organizers added *Les Autres* competitions—literally "the others"—a category that included athletes with "locomotor disabilities" such as short stature, multiple sclerosis, and limb length differences.[38] Athletes with intellectual impairments first took part in the 1996 Paralympic Summer Games in Atlanta.

For most of that time, athletes were grouped by impairment type, and each impairment type was governed by a separate organization (Table 3.1). This presented administrative battles, as well as multiple and often conflicting classification systems. By the early 1980s competitions had become "unwieldy," critics observed.[39] "We had way too many classes," recalls Dr. Robert Steadward, who began coaching Para sport in 1967 and later served as the IPC's founding president (1989–2001). "For the 100-meter race in track and field, there might be 17 to 20 classes, which was not very logical. No one understood it. The media did not understand it. I did not understand it. And most fans did not understand it. So the classification system not only had to become functional and sport-oriented, but we also had to reduce the number of classes."[40]

At the time, classification was conducted according to a "medical model," explains Steadward. The process in the 1970s and 1980s involved "strictly an anatomical and neurological evaluation and it

Table 3.1 Para Sport Organizations

Original Name	Later Names	Notes
International Stoke Mandeville Games Committee	International Stoke Mandeville Games Federation (ISMWSF), 1972	Merged with ISOD in 2002 to become International Wheelchair and Amputee Sports Federation (IWAS)
International Sports Organization for the Disabled (ISOD), 1964		Initially for athletes not included in the International Stoke Mandeville Games, such as amputee athletes and athletes with visual and locomotor impairments. ISOD merged with ISMWSF in 2002 to become International Wheelchair and Amputee Sports Federation (IWAS)
Sports and Leisure Group of the International Cerebral Palsy Society, 1969	Cerebral Palsy International Sports and Recreation Association (CPISRA), 1978	First CPISRA World Games, 1984
International Blind Sports Association, 1981	International Blind Sports Federation, 2002	
International Coordinating Committee Sports for the Disabled (ICC), 1982	International Paralympic Committee (IPC), 1989	The IPC replaced the ICC after the conclusion of the 1992 Paralympic Games in Barcelona.
International Sports Federation for Persons with Mental Handicap, 1986	International Sports Federation for Persons with Intellectual Disability, 1994 Virtus: World Intellectual Impairment Sport, 2019	

had nothing to do with the technical skills of the sport. Most of the people on the evaluation team were medical doctors who certainly understood the medical aspects but did not understand the technical elements within each sport."[41]

This deficiency was particularly evident in wheelchair basketball, a sport that developed at American rehabilitation hospitals during World War II for patients with spinal cord injuries. Men began competing at the 1960 Stoke Mandeville Games, joined by women's teams in 1968. In 1961, competitors were placed into one of two classes based on the severity of their impairment: the "complete division" for those with "total spinal cord injuries"; and the "incomplete division" for those with partial damage to the spinal cord or who had been affected by polio.[42] Yet the growth of the sport, the subsequent participation of athletes with different impairments, and the organizational difficulties associated with staging two separate competitions inspired change.

To combine the "complete" and "incomplete" divisions, the International Stoke Mandeville Games Federation introduced a system in 1966 that assigned each basketball player a point value based on manual muscle tests of his upper and lower extremities, and trunk balance. Initially, athletes were assigned 1, 2, or 3 points—with higher values associated with higher degrees of functioning. These points were then factored into a "team balance rule," in which the total points of the five players on the court could not exceed 15. Requiring teams to integrate athletes with varying degrees of activity limitation created a more inclusive sport, but athletes and coaches grew increasingly frustrated with the classification process.[43]

Much of the frustration was with the medical classifiers, who had "little affinity for or knowledge of the sport of wheelchair basketball," wrote Philip Craven, a Para athlete and the second IPC president (2001–2017).[44] The examiners did not consider how the athletes' impairments affected their ability to perform the different skills required for basketball. Players were frequently misclassified and had difficulty understanding the tests. They also found the process degrading, as Armand "Tip" Thiboutot, an accomplished player and a high-ranking official, articulated:

Having been pricked by pins, probed by the examiner's fingers, and ultimately dehumanized, the player is finally classified, almost always after being both prone and supine on an examination table, subject to the analytic gaze of the classifiers, who are usually standing, literally looking down on the player. Let's face facts: The player here has been reduced to a patient . . . the basketball player has been deathleticized."[45]

Despite these objections, the International Stoke Mandeville Games Federation, which governed wheelchair sports at the time, insisted on classifying players according to a medical model throughout the 1960s and 1970s.

In 1982, Horst Strohkendl, a coach and sport scientist, introduced a new "functional player-classification" model for wheelchair basketball. Strohkendl kept both the point system and the "team balance" concept, but integrated "medical, biomechanical (kinesiological), and specific technical information derived from both research and past experience," as he explained. Strohkendl also "listened attentively to the complaints of the players."[46] The result was to consider an athlete's degree of activity limitation relative to the sport's required skills. Accordingly, a "player's balance, or lack thereof, while shooting, rebounding, and pushing the wheelchair on the basketball court became more important than the muscle tests derived on a medical examination table."[47] Strohkendl's formula has been adapted over the years—at the 2020 Paralympic Games, players were graded from 1 to 4.5 points and total points on the court could not exceed 14—yet his basic consideration of the players as athletes, and not patients, is now championed by the IPC and has inspired classification in other sports.

In addition to Strohkendl's innovation, the early 1980s brought several changes to Para sport. In 1982, representatives from organizations for athletes with spinal injuries, cerebral palsy, amputations, and visual impairments came together to create the International Coordinating Committee Sports for the Disabled. But as Steadward recounts, "it was not a democratically developed organization" and was "structured by disability and not by sport." He therefore wrote and circulated an "extensive document" calling for the reorganization of Para sport, which delegates discussed at a pivotal 1987 meeting in Arnhem. There, they decided to form a global governing body that would eventually become the IPC, ushering in what Steadward calls the Paralympic movement's "modern era."

Participants in the 1987 Arnhem Seminar passed twenty-three resolutions, including the need to reduce the number of sport classes, "to implement a functional classification system," and to develop a system that would be structured "by sport and not by disability."[48] Almost immediately, Paralympic officials began pushing International Federations to create new classification systems. Six sports introduced these systems at the 1992 Games in Barcelona, where physicians, therapists, athletes, and coaches offered their expert opinions on the construction of sport classes.

Toward Research-Based Classification

Opinions weren't good enough for Dr. Sean Tweedy, who heads the University of Queensland's IPC Classification Research and Development Centre.[49] Tweedy began working in Para sport in 1983 and became a senior international classifier for track and field in 1994, but he found the process too subjective. He was particularly troubled by instances in which he had to tell athletes that they had been previously misclassified and would be competing in a class reserved for those with less severe impairments. This "meant that the athlete would have significantly reduced prospects of success," he explains. "So, the aim of the conversation was to ensure they had an opportunity to ask questions, that they felt heard, and that the reasons for the decision were fully explained." Although the process was as "good as it could be, there were gaps in the science such that when athletes disagreed with their classification and asked, 'Why are you putting me in this class?' the real answer was simply 'Because I said so.'"[50]

It was clear to Tweedy that classifiers needed a strong foundation on which to make their decisions. As he wrote in 2002, "It is critical that the definition and purpose have a sound scientific and taxonomic basis and are articulated using language and definitions that are unambiguous and internationally recognized." He therefore called for research that would "contribute to the refinement, validation, and clarification of disability sport classification in a rigorous, logical fashion."[51]

The IPC listened, and in 2003 its Governing Board recommended the development of universal classification requirements. The subsequent IPC Classification Code and International Standards include principles and guidelines consistent with the taxonomy and terminology of the World Health Organization's International Classification of Functioning, Disability and Health, as well as policies and procedures for the conduct of classification, and classifier certification. Each International Federation must comply with the IPC Code to publish classification rules and regulations for their respective sports.[52]

The IPC and its International Federations have worked hard to coordinate, justify, and validate their classification systems. At the core of each system is a three-step process: (1) determining the eligible impairment types for each sport; (2) determining the minimum impairment criteria athletes must meet to compete for each eligible impairment type; and (3) determining the number of sport classes for each eligible impairment type and the criteria for classification.

Eligible Impairment Types

First, International Federations select which impairment types are eligible to compete in their sport. Altogether, athletes with ten different impairment types compete in Paralympic sports. This includes intellectual impairments, visual impairments, and physical impairments, of which there are eight subcategories. Three of the physical subcategories are related to body structure, including limb deficiency, limb length difference, and short stature. The other five physical impairment types have to do with body function, including strength, range of motion, hypertonia (increased muscle tension), ataxia (impaired balance and muscle coordination), and athetosis (involuntary and unbalanced movements) (Tables 3.2 and 3.3).

Table 3.2 Sports at the 2020 Tokyo Summer Paralympic Games

Sport	Governing Organization	Physical Impairments	Visual Impairments	Intellectual Impairments
Archery	World Archery	★	★	
Badminton	Badminton World Federation	★		
Boccia	Boccia International Sports Federation	★		
Canoe	International Canoe Federation	★		
Cycling	International Cycling Union	★	★	
Equestrian	International Equestrian Federation	★	★	
Football, 5-a-side	International Blind Sport Association		★	
Goalball	International Blind Sport Association		★	
Judo	International Blind Sport Association		★	
Para athletics	World Para Athletics (IPC)	★	★	★
Para powerlifting	World Para Powerlifting (IPC)	★		
Para swimming	World Para Swimming (IPC)	★	★	★
Rowing	International Rowing Federation	★	★	
Shooting Para sport	World Shooting Para Sport (IPC)	★		
Table tennis	International Table Tennis Federation	★		★
Taekwondo	World Taekwondo	★		
Triathlon	International Triathlon Union	★	★	
Volleyball	World Para Volley	★		
Wheelchair basketball	International Wheelchair Basketball Federation	★		
Wheelchair fencing	International Wheelchair and Amputee Sports Federation	★		
Wheelchair rugby	International Wheelchair Rugby Federation	★		
Wheelchair tennis	International Tennis Federation	★		

Adapted from Sean M. Tweedy, Mark J. Connick, and Emma M. Beckman, "Applying Scientific Principles to Enhance Paralympic Classification Now and in the Future: A Research Primer for Rehabilitation Specialists," *Physical Medicine and Rehabilitation Clinics of North America* 29 (2018): 315.

Table 3.3 Sports in the 2018 PyeongChang Winter Paralympic Games

Sport	Governing Organization	Physical Impairments	Visual Impairments	Intellectual Impairments
Para alpine skiing	World Para Alpine Skiing (IPC)	★	★	
Para biathlon	World Para Nordic Skiing (IPC)	★	★	
Para cross-country skiing	World Para Nordic Skiing (IPC)	★	★	
Para ice hockey	World Para Ice Hockey (IPC)	★		
Para snowboard	World Para Snowboard (IPC)	★		
Wheelchair curling	World Curling Federation	★		

Adapted from Sean M. Tweedy, Mark J. Connick, and Emma M. Beckman, "Applying Scientific Principles to Enhance Paralympic Classification Now and in the Future: A Research Primer for Rehabilitation Specialists," *Physical Medicine and Rehabilitation Clinics of North America* 29 (2018): 315.

Only athletics and swimming offer competitions for all ten impairment types. Several sports are impairment-specific. Para judo, five-a-side soccer, and goalball (Figures 3.2 and 3.3), as examples, are only open to athletes with visual impairments, while Para powerlifting and Para ice hockey are only open to athletes with physical impairments that affect the lower limbs or hips (Figure 3.4). With twenty-two sports at the 2020 Summer Paralympic Games in Tokyo and six at the 2022 Winter Games in Beijing, there are multiple and often complex classification systems at play.

The IPC's Code stipulates that the athlete's impairment must be permanent and will not "resolve in the foreseeable future regardless of physical training rehabilitation or other therapeutic interventions."[53] This disqualified American swimmer Victoria Arlen, who officials

Figure 3.2 Australian Goalball Player Rob Crestani Shoots at the 2000 Sydney Paralympic Games Match, Australian Paralympic Committee/ Australian Sports Commission. Wikimedia Commons.

Figure 3.3 Sweden vs. Finland Women's Goalball 2012 Paralympics.
Wikimedia Commons.

Figure 3.4 Australian Powerlifter Richard Nicholson during 2000 Sydney
Paralympic Games Competition, Australian Paralympic Committee/Australian
Sports Commission. Wikimedia Commons.

decided had "failed to provide conclusive evidence of a permanent eligible impairment."[54] Although a rare autoimmune condition had paralyzed Arlen from the waist down, a panel of experts predicted she might eventually regain the use of her legs, and in fact she did. But after devoting years of her life to Para sport, the IPC's decision left her "heartbroken."[55]

The IPC Code also stipulates that the impairment must be relevant to the athlete's event. An athlete whose arm has been amputated below the elbow, for instance, would be eligible for Para swimming events. She would also be eligible for track events, but only in sprints of 400 meters or less, because the amputation site affects her use of the starting blocks. That same athlete is *ineligible* for Para track races greater than 400 meters because she would begin in a standing position.[56]

To be considered for Paralympic competition, an athlete submits a medical diagnostic form, completed by a physician, psychologist, or ophthalmologist, depending on the athlete's impairment type. The form is "absolutely essential to the process," explains Dr. Emma Beckman, a classifier and leading researcher at Queensland's IPC Classification Research and Development Centre. "This is key to answering the first important question of classification: does the athlete have an eligible impairment?"[57]

Minimum Impairment Criteria

Once an International Federation determines its impairment types, the second step is to set the minimum impairment criteria athletes must meet to compete in Para sport, or, in IPC-speak, "how severe an eligible impairment must be for an athlete to be considered eligible."[58] When World Para Swimming disqualified Ian Silverman, it was because classifiers believed he failed to meet those criteria. For Para judo, the International Blind Sports Federation stipulates that athletes must have less than 10 percent visual acuity or a visual field restricted to 40 degrees in diameter. Athletes with intellectual impairments must submit proof that their Intelligence Quotient is below 75, that they acquired their impairment before age eighteen, and that the impairment causes "significant limitations in adaptive behavior in conceptual, social, and practical adaptive skills," as outlined by the International Sports Federation for Persons with Intellectual Disability.[59]

Although International Federations have a degree of autonomy when regulating these criteria, their general principles must align with the IPC Code. The International Wheelchair Basketball Federation learned this the hard way in the lead-up to the 2020 Tokyo Games. The IPC threatened to drop the sport from the program because the basketball federation included the participation of athletes outside the IPC's ten recognized impairment groups. "If you have arthritis, you could play wheelchair basketball," railed IPC chief marketing and communications officer Craig Spence. "If you have a bad knee, you could play wheelchair basketball."[60]

While Spence raises important points about who is "impaired" enough for Para sport, he also downplays, if not belittles, athletes with significant impairments. Annabelle Lindsay's knee easily and frequently dislocates, making it impossible for her to participate in most "able-bodied" sports. "There is nothing I can do," explained the Australian basketball star:

> If I lift weights, it would have to be sitting down. I can't even go for long walks. I can only play adaptive sports—and not even all adaptive sports. Rowing would be too painful, my knee would just dislocate. It would have to be wheelchair-specific sports. But now they've even said, "sorry, you're not disabled enough to play in the Paralympics" but I'm definitely not able-enough to play able-bodied sports, so what do they expect me to do? Being able to play sport should just be a fundamental right that people have.[61]

Regulating impairment, as with any sporting regulation, is as much about "necessary discrimination" as it is about inclusion.

Sport Classes

The next step is for International Federations to settle on the number of sport classes and how to distinguish between them. Researchers are currently working on "evidence-based classification systems" that are founded on scientifically valid and reliable data.[62] Until those systems are in place, classification is done following the IPC's "Model of Best Practice."[63] This model varies from one sport to the next, but in effect best practice dictates that based on comprehensive data and repeated analyses, experts from each sport create "class profiles" that define the impairments and activities typical of athletes in that class. It is up to the classification panel, which assesses each athlete before competition, to match an athlete's specific limitations to the right profile.

None of this comes easily. Impairment exists on a "continuum," Beckman explains. The "tricky part is to find 'cut points'" on that continuum—sites at which researchers can justify separating one class from another. Some classes, such as those for athletes of short stature, are straightforward, but it gets trickier when classes include athletes with multiple impairment types. If an athlete is especially dominant in her sport, is it because she's simply talented, because she's been misclassified, or because her class includes a spectrum of activity limitations that is too broad?

It is also tricky to balance decisions about sport classes with practical concerns. If a class includes a wide range of activity limitations, those with greater limitations will be disadvantaged in competition. But the creation of many narrower classes presents a challenge to event organizers. There could be too many events to schedule in a reasonable time frame, or too few competitors to make each class viable. Excessive classes can also jeopardize "the sense of elite," says Tweedy. With thirty different 100-meter gold medalists in Paralympic track and field (sixteen for men and fourteen for women), spectators may fail to appreciate the magnitude of each.[64] In an already complex sport, added complexity may turn away fans, sponsors, and media coverage. In point of fact, researchers note that "one of the most profound hurdles in attempting to improve public understanding" of Para sport is the "particularly complex" classification system.[65]

The Classification Process

Classification takes place before an approved competition. The classification panel consists of at least two accredited classifiers: a medical classifier, typically a physical therapist, physician, physician's assistant, or occupational therapist; and a technical classifier who has expertise in the sport, such as a coach, scientist, or physical educator. The panel begins by certifying that the athlete has an eligible impairment type and that the impairment is severe enough to meet the minimum impairment criteria. This involves reviewing medical documents and detailed histories of training and performance. It also involves simple tests that might include measurements of stature or limb difference, manual muscle testing to ascertain impaired strength or stiffness, techniques to assess range of movement, and novel motor tasks.[66]

Next, the athlete performs sport-specific activities. Dr. Jonna Belanger is a technical classifier for USA Track and Field. As she explains, "We put them on the track and ask the runners to run while we observe. We ask throwers to perform from either their throwing chairs or the throwing ring."[67] The final step in the classification process is for the panel to observe athletes during competition to ensure their performances are consistent with their class allocation.

When in doubt about an athlete's class allocation, Belanger explains, "We will always disadvantage the athlete before we disadvantage an entire class." They assign the athlete to the class of athletes with lower degrees of activity limitation, "because if not," Belanger continues, "then we're basically picking the athletes who are going to win."[68] When misclassification is charged or suspected, there is a rigorous review and appeal process.

Classification can be tough on athletes. An explosion during her military service in Afghanistan left Shawn Morelli with brain trauma, loss of vision in her left eye, and severe damage to her spine and neck. She began cycling as a form of rehabilitation and, before long, she recalls, she learned she "was fast." She was right. Morelli quickly became one of the top athletes in the sport. Competing in the C4 class, reserved for athletes with impairments in one or two limbs, she has won, among her multiple titles, two gold medals at the 2016 Paralympic Games in Rio, and gold and silver in Tokyo 2020. But the tests she endures before competition can be "painful and exhausting" and often trigger muscle spasms that fatigue her before competition. The process is also "stressful," Morelli explains. One misstep in the process could assign her to the wrong class, where she would race against cyclists with more or less limitation.[69]

British wheelchair racer Hannah Cockroft has decried the "horrible tests to prove I am what I claim to be" (Figure 3.5). As the five-time T34 Paralympic champion and holder of five world records told

Figure 3.5 Hannah Cockroft Wins T34 100m Qualifying Heat, 2012 London Paralympic Games. Wikimedia Commons.

the BBC, "I've had all the scans . . . MRI scans, CAT scans. I think my worst one was I had to have electrodes attached to my spine and then electric shocks sent up and down my legs to see which nerves worked—that pain was sickening." The T34 class is specific to athletes who race in wheelchairs due to cerebral palsy or traumatic brain injury. Cockroft understands the need for repeated testing "with cases where a disability is varying, but my disability is pretty much the same every day."[70]

Nevertheless, most athletes must go through the classification process repeatedly. This verifies their class allocation, as some impairments can improve or worsen over time. Repeated classification also provides confidence that previous decisions, as well as the system itself, are consistent. In other instances, the system may change, which can present additional problems.

Reclassification

The Code "was never designed to be a document that stood still," the IPC maintains. "As classification and the Paralympic Movement developed, so would the ideas that would form rules, regulations and policies in the future."[71] It is a living document that adapts to new knowledge, innovation, and times. This is a good thing. In theory. In practice, reclassification, whether in the form of a new code or allocating an athlete to a different class, can have significant if unintended consequences.

Australian swimmer Jacqueline Freney originally competed in the S8 class, where she won three bronze medals at the 2008 Paralympic Games in Beijing. However, Australian officials believed that she had been misassigned and that her degree of activity limitation was greater than the classifiers' original assessment. Following a process of appeal and review, they reassigned Freney to the S7 class. She went on to win eight gold medals at the 2012 Games in London. Good for Freney, but not necessarily for the other S7 competitors.

It was the opposite situation for American Mallory Weggemann, who, swimming in the S7 class, won eight golds and one silver medal at the Beijing Games before setting nine world records at the 2010 World Championships. Just before the start of the London Games, where she had her sights set on nine golds, the IPC reclassified her to S8. At the time, she described herself on the "borderline of the S6 and the S7 category" because she is "a T10 complete paraplegic"— meaning she sustained an injury at her tenth thoracic vertebra (T10), located at mid-spine. Although Weggemann says she has "no feeling or movement from my belly button down," the reclassification forced her to compete "against people who are bi-lateral double amputees below the knee, who have, from below the knee up, full function," as she explained. "It's like competing against seven athletes who are doping, in a sense, when you're at that big of a disadvantage."[72]

In other cases, the athletes' impairments or activity limitations don't change, but the system does, as it did in 2018 with World Para Swimming. The year before that, World Para Athletics, the International Federation for track and field, announced new testing procedures and several organizational adjustments. Notably, the T/F 42–44 classes, which previously included track (T) and field (F) athletes with prosthetic lower limbs, would be reserved only for athletes with lower limb impairments who run and jump on their anatomical legs. Athletes who run or jump with prosthetic lower limbs were reassigned into four new categories: T/F 61 to T/F 64. As a result, all previous world and regional records in the T/F 42–44 classes were "archived" but essentially erased, a devastating blow to many world-class athletes.[73]

Intentional Misrepresentation

Intentional misrepresentation is among the reasons that Para sport officials remain vigilant about class assignments. The term refers to athletes who try to game the system by exaggerating their limitations during classification. They tank the tests—acting as though they are weaker, less coordinated, or have a lower range of motion or slower reaction times than they really do. This allows them to compete in a class for those with more severe impairments and increases their chances for success. It is a serious problem, one that occurs regularly enough for the IPC to decree that anyone found guilty risks a two-year suspension from Para sport. A second offense results in "a lifetime period of ineligibility," according to the IPC Code.[74]

Intentional misrepresentation seems to be most prevalent among athletes with neurological conditions. It is impossible, after all, to fake short stature or a missing limb. Athletes take different tactics. One Australian swimmer confided to reporters that just before classification, several of her teammates would "go for long runs, or bike rides, to increase how great the effect of their disability is. . . . Others would take cold showers. I've heard of one person saying they were classified in a colder climate so they were chucked in the snow a few minutes before their classification so they would tighten up." The cold water and snow temporarily increased spasticity, which classifiers assess. The swimmer continued to explain she "had a coach who once told me before my classification process that I should think about getting a stress ball and squeezing it so I would appear fatigued. It's common." Other witnesses allege that they have seen competitors take muscle relaxants or, conversely, strap or bandage their legs to stiffen their muscles.[75] All of this is meant to make the athletes seem more impaired during the classification stage so that they can compete in a favorable class.

Charges of intentional misrepresentation have been continually levied against Ukrainian swimmer Denys Dubrov, who, as a member of his country's "able-bodied" team, set a national record in the 200-meter individual medley. In April 2014, he competed as an "able-bodied" swimmer in a Ukrainian Championship. In July of that same year, Dubrov turned up at the IPC European Championship, where he won three medals. There was no explanation about how, in just four months, he came to meet World Para Swimming's minimum impairment criteria. He amassed even more success at the 2015 IPC World Championships and the 2016 Paralympics, where he won eight medals, including three gold.

If there is any cheating going on, it might be bigger than Dubrov. For decades, Ukraine has finished among the top six countries in the Paralympic medal count, "despite consistently being ranked among the poorest countries in Europe and cited by the United Nations as a difficult home for people with disabilities," the *New York Times* reports.[76] This has raised some eyebrows. As journalist Deirdra Dionne asks, "Is Ukraine just being strategic and implementing talent identification to maximize the country's medal count, or are they manipulating the system to win more medals?"[77] In the 200-meter individual medley at the 2016 Rio games, defending champion Benoit Huot finished fourth behind Ukrainians Dubrov, Maksym Krypak, and Dmytro Vanzenko. "I'm number one in the world—other than Ukraine," Huot joked. "I just hope it's a fair system."[78] He was being diplomatic. Others, including Ian Silverman, do not mince words. Dubrov "is faking a disability and playing with the world's perception of the Paralympic movement," Silverman wrote to the IPC. He is guilty of "the most outrageous fraud the Paralympic Movement has ever seen."[79]

If Silverman is right, Dubrov would have a hard time taking the "most outrageous fraud" title away from the 2000 Spanish Paralympic basketball team that won gold in the intellectual impairment class. Shortly after the Games, journalist and team member Carlos Ribagorda revealed that ten of the twelve athletes on the team were not impaired. The Spanish Federation of Sportspeople with Intellectually Disabilities, he disclosed, had deliberately signed unimpaired athletes to "win medals and gain more sponsorship." Upon further investigation, the IPC estimated that in 2000, 69 percent of medals awarded to athletes claiming intellectual impairments came "from various countries whose [International Sports Federation for Persons with Intellectual Disability] registration forms did not meet the proper requirements of the eligibility verification process."[80] According to

Steadward, who served as the IPC president at that time, "they did not use certified clinical psychologists or psychiatrists to evaluate the athletes. When we looked at the entry forms, they were signed by moms and dads, teachers, principals, and social workers, and the like."[81] The 2000 debacle had serious ramifications, not least of which was the IPC's suspension of events for athletes with intellectual impairments until 2012.

There can be extra pressure to intentionally misrepresent one's activity limitation in programs where funding is dependent upon performance. The Australian Sports Commission and its Winning Edge Program, for example, provide funds to Swimming Australia. Failure to produce top performances could mean a reduction or loss of those funds. When the mother of a Paralympic swimmer raised her concerns about intentional misrepresentation, a Swimming Australia official responded in an email:

> Should you continue making such communications you run the risk of breaching IPC Classification code and IPC swimming rules, policies and procedures which may lead to [the Australian Paralympic Committee] or IPC Swimming taking direct action against your daughter thus jeopardizing [her] future in Paralympic Swimming.[82]

Shut up or your daughter is off the team.

Elite British sport is similarly dependent upon UK Sport's "no compromise" approach to funding, where athletes must win for their sport to receive financial support from the government.[83] As a result, there have been allegations that UK Athletics knowingly misclassified athletes to win medals and maintain its budget. It was for this reason that British sprinter Bethany Woodward, who won silver in the 2012 Paralympic T37 200-meter race, withdrew from the Paralympic squad. As she told reporters, "I represented my country for a long time but if I can't compete like I used to compete, because they've brought in people who are not like me in terms of disability, what's the point?" She later returned a 400-meter relay medal because, she claimed, one of her teammates gave "us an unfair advantage . . . I feel like we won a medal I don't believe was true."[84]

Intentional misrepresentation is among the biggest threats to Paralympic legitimacy. Six-time Paralympian Stephen Miller thinks the problem cost him "a few medals" during his career.[85] Every day the IPC receives cheating allegations. In advance of the 2016 Paralympic Games, IPC officials reviewed the files of more than eighty athletes across six sports from twenty-four countries based on accusations of intentional misrepresentation. In each instance, IPC officials were unable to find evidence of cheating "beyond reasonable doubt."[86] Even so, the practice undoubtedly goes on and mars the reputation of Para sport.

An Impairment Advantage?

Classification creates "protected classes" within Para sport by discriminating between athletes' degrees of activity limitation. At the same time, the minimum impairment criteria protect Para athletes by discriminating them from their "able-bodied" peers. Yet there are no regulations that work in reverse. Between 1904 and 2016, at least twenty-eight athletes with Paralympic-worthy physical or visual impairments competed in the Olympic Games.[87] The first was probably George Eyser, who lost his left leg in a train accident. At the 1904 Games in St. Louis, competing with a prosthesis made of either wood or cork (accounts vary), Eyser collected six Olympic medals in gymnastics, including three golds in the parallel bars, long horse vault, and 25-foot rope climb; two silvers in pommel horse and four-event all around; and bronze in the horizontal bar (Figure 3.6).[88]

Olivér Halassy, despite losing his left foot in childhood, won multiple swimming competitions and helped the Hungarian water polo team win Olympic gold in 1932 and 1943 (Figure 3.7). At age twenty-three, Danish equestrian Lis Hartel contracted polio that left her partially paralyzed below the knees. She went on to win a silver medal in individual dressage at the 1952 Olympic Games—where she was, incidentally, the first woman to ride for an Olympic equestrian team

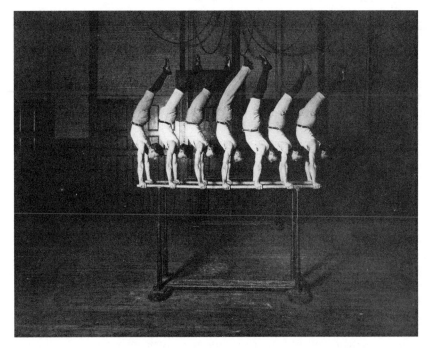

Figure 3.6 George Eyser, center, with Concodia Turnverein, 1907, Louis Melsheimer. Missouri History Museum.

Figure 3.7 1932 Olympic Gold Medal Hungarian Water Polo Team, Los Angeles. Olivér Halassy on far right. Wikimedia Commons.

Figure 3.8 Lis Hartel and Jubilee at the 1952 Summer Olympics in Helsinki. Wikimedia Commons.

(Figure 3.8). Until 1952, Olympic dressage was restricted not only to men but to military men.

No one claims that Eyser, Halassy, or Hartel had an unfair advantage over their "able-bodied" competitors, but there are other cases that are less clear-cut, particularly when it comes to the use of assistive technologies. Wheelchair athletes Neroli Fairhall and Paola Fantato competed in the Olympic archery contests of 1984 and 1996, respectively. Able to lock their chairs into place, did the women benefit from stabilizing their lower bodies in ways that other competitors could not? Even more controversial is the use of state-of-the-art prostheses in track and field, best exemplified by champion sprinter Oscar Pistorius, since convicted of murder. But before he was a murderer he was "the fastest man on no legs."

The nickname lacks accuracy. Pistorius has biological legs—they just end below his knees due to a bilateral transtibial amputation. Pistorius also has prosthetic legs, including his state-of-the-art

Figure 3.9 Oscar Pistorius, 2011 World Championships Athletics in Daegu, South Korea, Erik van Leeuwen. Wikimedia Commons.

"Cheetah Flex-Foot" racing legs—and that is what made his Olympic bid so controversial (Figure 3.9).

Pistorius quickly rose to the top of Para athletics in the 100-, 200-, and 400-meter races, with times that began to rival elite "able-bodied" sprinters. In 2007, the IAAF, anticipating Pistorius's petition to compete in its events, amended its rules to ban the use of "any technical device that incorporates springs, wheels or any other element that provides a user with an advantage over another athlete not using such a device."[89] This effectively disqualified Pistorius, who was unable to race without the use of a "technical device."

Pistorius appealed to the CAS, claiming, in part, that the IAAF had failed to meet its "burden of proof" that on the "balance of probability," the blades gave him an "overall advantage" over other runners.[90] In other words, it was more likely than not that the blades enhanced his performance. The CAS arbitrators ultimately sided in his favor, citing flawed and contradictory studies of Pistorius's performance, and ruled that he could compete in IAAF-sanctioned events "with immediate effect."[91] Although Pistorius failed to meet the qualifying times for the 2008 Summer Olympic Games, he did compete in London 2012, where he advanced to the 400m semifinals and finished eighth place as a member of South Africa's 4 × 400m relay. Two weeks later, he added three Paralympic gold medals to his collection.

Whereas the IAAF accepted the burden of proof in the Pistorius case, the federation later abdicated this responsibility by again amending its competition rules in 2015. The updated competition rule prohibited "the use of any mechanical aid, *unless the athlete can establish on the balance of probabilities* that the use of such an aid would not provide him with an overall competitive advantage over an athlete not using such aid."[92] This meant that the athlete must bear the brunt of all costs and efforts to prove a level playing field.

In 2019, the IAAF concluded that American sprinter Blake Leeper had failed to meet "his burden of proof" and denied his application to race using his "passive-elastic carbon-fiber running specific-protheses." Hoping to compete in the 2020 Olympic Games, Leeper appealed to the CAS to declare the rule "invalid." Among his principal arguments, he charged discrimination because "able-bodied athletes (who do not require a prosthetic) are not subject to any pre-eligibility burden to prove that they do not have any type of overall competitive advantage before they are allowed to compete."[93] This violated the United Nations Convention on the Rights of Persons with Disabilities (CRPD), he continued. Specifically, Article 30(5) of the convention reads:

> With a view to enabling persons with disabilities to participate on an equal basis with others in recreational, leisure and sporting activities, States Parties shall take appropriate measures: (a) To encourage and promote the participation, to the fullest extent possible, of persons with disabilities in mainstream sporting activities at all levels.

Leeper's prostheses, the lawyers argued, "did not provide him with any competitive advantage over able-bodied competitors, and instead merely provided him with the opportunity to be able to compete with them."[94] Allowing him to compete in IAAF events aligned with the principles of the CRPD.

The federation remonstrated. "The IAAF is not a State Party, nor a public authority exercising state powers. It is a private body exercising private, contractual powers. As such, the CRPD does not apply to it."[95] Just as in Caster Semenya's case, the federation claimed "sport exceptionalism" to exempt itself from international human rights accords, and again, the CAS panel agreed (Chapter 2).

The CAS panel also agreed with the IAAF's assessment that Leeper's prostheses granted him an unfair advantage, specifically because they made him taller than he would have been on biological legs.[96] While this left him ineligible for the 2020 Olympic Games, the panel "partially upheld" his appeal by finding that placing the "burden of proof" on the athlete was "indirectly discriminatory" because it only affected athletes with impairments.[97] Consequently, to exclude an athlete's prosthesis, the IAAF (now World Athletics) must assume responsibility for proving that prosthesis provides an unfair advantage.

The challenges issued by Pistorius, Leeper, and others raise important questions. As one interdisciplinary research team asked, "ought we to encourage direct competition between people who do and do not require enabling technology in elite-level sports events such as the Olympics?[98] This would, after all, correspond with the CRPD and other nondiscriminatory covenants. "Or does pursuit of such in-clusion simply betray a tacit value judgement—that the Paralympics are 'second best' and that in order to be 'truly' elite, an athlete must compete at the Olympics?"[99] Hierarchy is the unfortunate partner of difference, and separate is rarely equal. Critics who oppose inte-grated sport argue that the use of prostheses and other assistive devices is tantamount to "technology doping."[100] As explored in Chapter 4, however, just what doping means—and whether and how to regulate it—is up for debate.

4

Regulating Dope

Three days before the start of the 1998 Tour de France, French customs officials near the Belgium border stopped Willy Voet, an assistant with the Festina cycling team. In their search of Voet's car, officials found amphetamines, synthetic testosterone, growth hormones, and nearly 250 vials of recombinant erythropoietin (EPO), all banned in sport for their performance-enhancing potential. Voet was taken into custody where he eventually confessed: for years, he had supplied cyclists with prohibited drugs; team doctor Eric Rijkaert administered the drugs; all nine of the Festina cyclists on the Tour were doping; Festina team manager Bruno Roussel oversaw the entire operation.[1]

The dominoes fell quickly. Police raided Festina's offices in Lyon, where, in addition to more drugs, they discovered documents that detailed systematic doping programs for the team's riders. Under France's 1989 Anti-Drug Act, Roussel and Rijkaert were arrested and both admitted their complicity. Festina was expelled from the Tour de France, already in progress.

What has gone down in history as the "Festina Affair" was much bigger than one team. During the 1998 Tour, police raided the vehicles and hotel rooms of several teams, interrogated riders and their support staff, and subjected athletes to a battery of tests. In the end, seven teams were either disqualified, left under a cloud of suspicion, or quit in protest. Of the 189 riders who started the competition, only 98 finished.[2] Disillusioned spectators jeered and sponsors withdrew

Figure 4.1 "Tour de Doping," Protest Banner at the 2006 Tour de France. Wikimedia Commons.

their support as it became clear that doping on the Tour was both ubiquitous and largely unchecked by the International Cycling Union's (*Union Cycliste Internationale* or UCI), cycling's governing body (Figure 4.1).

★★★★★

The 1998 Tour de France was not just a blight on the UCI but also on the International Olympic Committee (IOC), which, since the 1970s, had assumed the policy leadership role in the anti-doping movement.[3] While the IOC devised its own set of anti-doping regulations, they were often at odds with the regulations of International Federations, sports leagues, national anti-doping agencies, and national governments. Certainly, the IOC wanted to protect the health of athletes and the spirit of fair play, but the most immediate concern following the Festina Affair was to protect its own interests. As scholar John

Hoberman argues, "doping was primarily a public relations problem that threatened lucrative television and corporate contracts."[4]

The IOC consequently convened the 1999 World Conference on Doping in Sport to establish a new agency that would direct a global anti-doping movement. The conference did not go as planned. Rather than agreeing to the IOC's terms, representatives from national governments pushed back by recalling recent misconduct within the Olympic movement. Barry McCaffrey, director of the United States' White House Office of National Drug Control Policy, excoriated the committee for its "alleged corruption, lack of accountability, and the failure of leadership." British Sports Minister Tony Banks concurred. "We support a totally transparent world anti-doping organization," he told conference attendees, "but the IOC should not be that agency."[5]

Instead, that agency became WADA—the World Anti-Doping Agency—which operates independently from the IOC, although the IOC provides half the agency's funding. The other half comes from participating national governments, and WADA's leadership is shared between Olympic and government representatives. The resulting structure is an "unusual, and possibly unique, hybrid intergovernmental-private body."[6]

A number of responsibilities fall under WADA's purview, including anti-doping research, education, the accreditation of laboratories, developing testing protocols, setting penalties for rule violations, annually updating the list of prohibited substances and methods, and protecting athletes' privacy and personal information. Most fundamental to WADA's mission, though, is the publication and periodic revision of the World Anti-Doping Code, "the core document that harmonizes anti-doping policies, rules and regulations within sport organizations and among public authorities around the world," as the agency explains.[7] Any International Federation that hopes to see its sport on the Olympic or Paralympic program, and any nation that intends to send a delegation to the Games must sign the Code.

It is tempting, and not altogether inaccurate, to recount the history of anti-doping efforts through a series of scandals. Indeed, write experts Ivan Waddington and Verner Møller, since the 1960s, "each crisis has been followed by demands for more testing and, when this fails, demands for yet more tests and further intensification."[8] This chapter offers a truncated account of that history, touching briefly on select moments that altered the course of the anti-doping movement. This leads to a discussion of the three criteria WADA considers when ruling on the permissibility of a particular substance or method: performance enhancement, health, and its relation to the "spirit of sport." In line with the other policies explored in this book, WADA's expressed intent is to protect the ethos of fair competition, to protect athlete's health and well-being, and to protect the image and interests of sport.

There are, of course, unintended consequences to WADA's policies, including the violation of athletes' rights, unhealthy behaviors and deceptive practices, and impeding the progress of sport and human potential. There are also questions about how effective the anti-doping movement is. WADA reported that in 2020, only 0.67 percent of tests resulted in "adverse analytical findings," its term for evidence of the use of a prohibited method or substance.[9] This percentage is not only suspiciously low, it also contradicts the existing research. The authors of a 2105 study estimated that somewhere between 14 and 39 percent of elite athletes dope, but the numbers vary according to sport.[10] One investigation determined that 46 percent of medal winners in international Nordic skiing returned at least one abnormal drug test.[11] In another survey, conducted on the condition of anonymity, 57 percent of elite track and field athletes admitted to using performance-enhancing substances.[12] And cycling competitions have long been "special hotbeds of doping," described Ludwig Prokop, a physician and IOC advisor.[13] All of this suggests that WADA's testing protocols fail to catch a significant number of dopers. With that in mind, as well as additional problems that plague the anti-doping movement, this chapter concludes with a review of possible alternatives to current protective policies.

Early Anti-Doping Regulations

Outside of sport, humans have always sought ways to overcome fatigue, increase endurance, heighten aggression, sharpen mental focus, calm down, boost strength, and induce general feelings of euphoria, well-being, and escape. It is no surprise that athletes seek these same benefits. Ancient competitors fiddled with their diets, availed themselves of sesame seeds, hallucinogenic mushrooms, various herbs and plants, and dabbled in organotherapy, that is, the consumption of human or animal organs (hearts, livers, brains, testicles, etc.).[14] Importantly, none of that was doping, but rather "ergogenic"—intended to enhance performance, stamina, or recovery. With no anti-doping rules in place, early athletes seeking to enhance performance committed no offense. As sport scholar John Gleaves explains, "doping is really a notion that only makes sense in modern sport."[15]

According to Gleaves, the first anti-doping efforts had to do with horses, not humans. Unscrupulous track-goers were known to slip a little something to unsuspecting equines—either to pep them up or slow them down. In the late 1800s and early 1900s, jockey clubs in Europe and North America began to formally ban the practice as newspaper headlines decried doping as "the evil of the turf."[16] The moral outrage expressed over drugging horses was not about performance enhancement or the health of the animals, but rather about the effects of doping on gambling, which, as discussed in Chapter 2, played a pivotal role in developing the rules of modern sport.[17] By 1912, there were methods in place to test horses' saliva for the presence of cocaine, heroin, and opium.[18]

When it came to humans, initial concerns focused exclusively on stimulants. Athletes, coaches, trainers, and physicians experimented with different substances to speed up the central nervous system: kola nuts, caffeine, ephedra, sugar cubes dipped in ether, purified oxygen, alcohol, belladonna, nitroglycerine, and cocaine, to name a few. A cocktail of brandy, egg whites, and strychnine sulfate propelled

Thomas Hicks across the finish line to take the gold medal in the 1904 Olympic marathon (only after officials disqualified top finisher Fred Lorz for riding eleven miles of the race in an automobile). After attending to Hicks, physician Charles Lucas declared that "the Marathon race, from a medical standpoint, demonstrated that drugs are of much benefit to the athletes."[19]

As athletes in other sports chased these same benefits, concern grew over the use of ergogenic substances, especially in amateur sport.[20] These concerns developed in tandem with anxieties about women athletes (Chapter 2). Dopers and gender-diverse competitors were both considered "cheats" that threatened the sanctity of sport and the Olympic movement, and protective policies flowed from that position.[21]

In 1928, the IAAF (now World Athletics) became the first international organization to ban doping, which it narrowly defined as "the use of any *stimulant* not normally employed to increase the power of action in athletic competition above the average." The penalty was suspension from sport "for a time or otherwise from further participation in amateur athletics under the jurisdiction of this Federation."[22]

Ten years later the IOC similarly pronounced, in somewhat more expansive language, that "the use of *drugs or artificial stimulants of any kind* must be condemned most strongly, and everyone who accepts or offers dope, no matter in what form, should not be allowed to participate in amateur meetings or in the Olympic Games."[23] Such decrees had little effect. As Danish physician Ove Bøje wrote in 1939, "There can be no doubt that stimulants are today widely used by athletes participating in competition; the record-breaking craze and the desire to satisfy an exacting public play a more and more prominent role, and take higher rank than the health of the competitors itself."[24] Even before the turn of the century, the public and commercial thirst for better sport encouraged risky behavior.

During this same time, drug companies began to manufacture and distribute amphetamines, or synthetic stimulants. First used to treat nasal congestion, by 1936, writes historian Nicolas Rasmussen, "almost any use was imaginable, since amphetamine had adrenaline-like effects on the circulation, respiration, and involuntary muscles, as well as the brain."[25] Several countries issued amphetamines to their troops during World War II to help fight fatigue, improve endurance, heighten aggression, and boost morale. This was accompanied by scientific studies that led to the drug's widespread use at war's end.

The proliferation of amphetamines in the postwar era was part of what physician Louis Lasagna termed the "pharmaceutical revolution," characterized by "extraordinary acceleration in the discovery, development, and delivery of chemicals used in the diagnosis, prevention, and treatment of human disease."[26] As international sport accrued greater sociopolitical significance in the Cold War era, athletes discovered that the therapeutic chemicals of the pharmaceutical revolution also enhanced performance.

Doping in sport continued largely unbridled until the death of Danish cyclist Knud Enemark Jensen at the 1960 Olympic Games. Although Møller has since debunked rumors that link amphetamines to Jensen's death, those rumors were nevertheless a catalyst for the modern anti-doping movement.[27] Concerned parties formed committees and held conferences to address the issue, but it was not until 1965 that scientists developed a test for amphetamine use, first trialed at that year's Tour of Britain.

At the same time, Belgium and France passed national anti-doping laws that were put to the test during the 1966 Tour de France. Under the direction of France's Ministry of Youth and Sports, physicians ambushed riders in their hotel rooms after one stage of the Tour. The physicians interviewed the athletes about their diets and medical treatments, searched their bodies for needle marks, and collected urine samples to be analyzed in a Paris laboratory. The following morning, cyclists registered their collective protest by walking their bikes for the first 50 meters of that day's stage while shouting, "No to pissing in the test tubes."[28] The protests continued later that year at the UCI World Championships, where several top finishers refused to be tested.

Those protests stalled after the death of English cyclist Tom Simpson at the 1967 Tour de France. During the thirteenth stage of the contest, Simpson fell near the summit of Mont Ventoux, the "Giant of Provence." Doctors pronounced him dead at a nearby hospital and an autopsy confirmed his use of amphetamines.[29] Critically, scholars note that amphetamines did not directly cause Simpson's death, but were one of a confluence of factors, including elevated temperatures, overexertion, dehydration, and his consumption of alcohol during the competition, as was the tradition of the time (see Chapter 1). Still, anti-doping crusaders propagandized his drug use for their burgeoning campaign.[30]

The IOC Takes Control

The high-profile cycling deaths of the 1960s, combined with other, less-publicized incidents, compelled the IOC to take more strident action in the war on dope. In 1967, the committee's nascent Medical Commission devised protocols for drug testing and analysis and officially banned five classifications of substances:

a) Sympathomimetic amines (e.g., amphetamine), ephedrine and similar substances.
b) Simulants of the central nervous system (strychnine) and analeptics.
c) Narcotics and analgesics (e.g., morphine), similar substances.
d) Anti-depressants (e.g., IMAO), imipramine and similar substances.
e) Major tranquilizers (e.g., Phenothiazine).[31]

The Medical Commission rolled out its first anti-doping procedures at the 1968 Winter Olympic Games in Grenoble, where no athletes tested positive for any of the banned substances. Later that year, at the Summer Games in Mexico City, modern pentathlete Hans-Gunnar Liljenwall had the distinction of being the first—and at the time only—Olympic athlete disqualified for doping. His drug of choice? Beer. Liljenwall drank to steady his nerves before the shooting competition, resulting in "too great an amount of alcohol" in his system, according to the commission.[32] The IOC stripped Liljenwall and his Swedish teammates of their bronze medal, and alcohol remained a prohibited substance for the next fifty years.

In the beginning, the IOC only banned substances for which there were proven tests for detection.[33] This left androgenic anabolic steroids off the list. A synthetic derivative of testosterone, steroids were initially developed in the 1930s to treat a variety of medical conditions. In the 1950s and 1960s, weightlifters and bodybuilders seized on steroids' tissue-building properties, and before long athletes in track and field, American football, and other sports did the same.[34]

Scientists finally developed tests to detect the use of steroids in 1973, prompting the IAAF to ban their use the following year. The IOC followed suit, and at the 1976 Summer Games in Montreal eight athletes tested positive, including seven male weightlifters and a female discus thrower.[35] In retrospect, it is astonishing that no athletes from the German Democratic Republic (GDR, or East Germany) were caught, and especially no women swimmers, who won eleven of the thirteen women's individual gold medals and set eight world records at the 1976 Games. Despite persistent suspicions that anabolic steroids were part of the national training program, it was only after the 1990 reunification of Germany that the world learned just how deep and systematic the doping was.[36]

The collapse of the GDR led to the discovery of extensive documents on the orchestrated doping of as many as 10,000 athletes, including adolescents, many of whom were only told that the drugs were "vitamins."[37] In concert with a finely tuned system of talent identification and development, along with cutting-edge scientific research and coaching, the small GDR became "the most successful sports nation in the final two decades of the Cold War," writes historian Mike Dennis.[38] In the years since, however, approximately 5 to 10 percent of the athletes who were part of the program have experienced long-term health problems, including emotional trauma, cancer, liver damage, heart disease, gynecological conditions, and congenital anomalies in their children.[39] In the absence of studies on the health and performance-enhancing effects of steroid use in athletes, anti-doping advocates cite evidence from the GDR as support for their mission.[40]

East Germany is not the only nation to be involved with state-sponsored doping, although it may have been the most effective. The history of sport either reveals or hints at multiple examples of "socially-organized doping," which scholars define as "networks of interaction used by a team or nation to enable doping."[41] It was through such programs that athletes began experimenting with techniques to increase their circulating red blood cells to improve aerobic capacity and performance. They initially achieved this through homologous transfusions, which use the blood of matched donors, and autologous transfusions, which remove and store athletes' own blood, allow their bodies to recover, and then reinfuse the stored blood before competition.[42] While the IOC condemned blood doping, it stopped short of banning it because, once again, there was no way to detect it.

The coaches of the 1984 US Olympic cycling team broke no formal policy, then, when they oversaw the blood-boosting of at least eight American riders. With limited time and operating out of a hotel room, the staff used the whole blood donated by the riders' friends and families. "I'm surprised nobody died," remarked cyclist Connie Carpenter, who refused the transfusion and went on to win gold in the women's road race.[43] Not only did the doped riders survive, but they helped end a seventy-two-year medal drought for US cycling, winning nine medals, four of which were gold.

The IOC's 1986 prohibition on blood doping marked three significant changes in anti-doping history. First, it effectively altered the definition of doping to include banned methods and techniques, in addition to substances. Second, a substance or method could now be banned without the existence of a test for detection. Third, it opened the door to blood testing, in addition to the standard urinalyses.

Athletes and advocates objected to giving blood samples, and it was not until the mid- to late 1990s that national and international sport organizations adopted the procedures. Yet most organizations, including the IOC, resisted blood tests "until the scientific, ethical, and legal issues are fully explored and resolved."[44] In the end, blood tests were not mandated across all sports until WADA's first Anti-Doping Code came into effect in 2004.

Another controversial change to testing protocols developed after the 1988 Olympic men's 100-meter race, perhaps "the dirtiest race in history."[45] There, Canada's Ben Johnson won in a world record time of 9.79 seconds. Two days later, the IOC stripped Johnson of his gold medal for his use of the anabolic steroid stanozolol. Johnson was not alone. Six of the eight finalists in that race were implicated, in one way or another, in doping throughout their careers. A subsequent *New York Times* investigation found that at least half of the 9,000 competitors at the 1988 Games used some type of performance-enhancing drug.[46]

At issue was that unlike Johnson, most athletes stopped using steroids before a competition, as sport historian Thomas Hunt details.[47] In a cycle practically perfected by the East Germans, athletes used steroids to train harder and more frequently, after which they would taper off in advance of testing. This was effective with other drugs as well, begging the question of whether to distinguish between enhanced performance and advanced training. Anti-doping crusaders shunned the distinction and instead raised the call for unannounced "out-of-competition testing" that could occur at any time and place outside of scheduled contests. At the time, most International Federations balked at the time and effort involved in the tests, as well as the invasion of athlete's private lives.[48]

As with blood testing, the first World Anti-Doping Code codified out-of-competition across all sports. WADA went a step further with its "whereabouts rule," which it claims is "one of the most powerful means of deterrence and detection of doping."[49] According to the 2021 Code, every three months, a select group of athletes must submit the following information to anti-doping authorities:

- Home address, email address and phone number
- An overnight accommodation address
- Regular activities, such as training, work, school, or university and their locations plus times they will be there
- Competition schedules, including when they are taking place and where
- A sixty-minute time slot for each day where they'll be available and accessible for testing

On any day of the year, control agents can appear during that sixty-minute window and test the athletes. Failing to file the required information or missing three tests constitutes an Anti-Doping Rule Violation, and three "Whereabouts Failures" in a twelve-month period result in a minimum one-year suspension from sport (Table 4.1).[50]

Table 4.1 World Anti-Doping Code 2021, Article 2, Anti-Doping Rule Violations

Article 2.1	Presence of a prohibited substance or its metabolites or markers in an Athlete's sample
Article 2.2	Use or attempted use by an Athlete of a prohibited substance or a prohibited method
Article 2.3	Evading, refusing, or failing to submit to sample collection by an Athlete
Article 2.4	Whereabouts failures by an Athlete
Article 2.5	Tampering or attempted tampering with any part of doping control by an Athlete or Other Person
Article 2.6	Possession of a prohibited substance or a prohibited method by an Athlete or Athlete Support Person
Article 2.7	Trafficking or attempted trafficking in any prohibited substance or prohibited method by an Athlete or Other Person
Article 2.8	Administration or attempted administration by an Athlete or Other Person to any Athlete in-competition of any prohibited substance or prohibited method or administration or attempted administration to any Athlete out-of-competition of any prohibited substance or any prohibited method that is prohibited out-of-competition.
Article 2.9	Complicity or attempted complicity by an Athlete or Other Person
Article 2.10	Prohibited Association by an Athlete or Other Person

Athletes, players' unions, and sports organizations have unsuccess-fully challenged the whereabouts rule on the grounds that it is dra-conian, onerous, and violates their human rights. Only one other group of people is "required regularly to report their whereabouts to the authorities," argues Waddington: "convicted criminals who have been released from prison early on parole, and criminals who are considered particularly dangerous, such as those who have been con-victed of sexual offences against children."[51] Tennis star Rafael Nadal articulated a similar position. "They make you feel like a criminal . . . It is not fair to have persecution like that."[52] For an athlete whose elite career spans twelve years, calculate Paul Dimeo and Verner Møller, that sixty-minute time slot for testing amounts to "a total of 4,380 hours, equaling 182 full days, which is more than six months" of the athlete's life spent waiting for the whereabouts testers.[53]

Supporters of the whereabouts rule point out that elite sport is "voluntary practice," and that athletes who object to its policies do not have to participate.[54] As high jump champion Blanka Vlašić ra-tionalized, "it is a price we need to pay for being at the top level in our job."[55] Engaging in elite sport means consenting to the rules of the game, but consent is a sticky term. For instance, philosopher Angela Schneider contends that "athletes living below the poverty line may well feel that they are coerced into giving consent" to the tests be-cause they feel they have no other choice.[56] For that matter, athletes may be coerced into doping, either by coaches, parents, sponsors, gov-ernment officials, or the culture of sport. How much could fifteen-year-old figure skater Kamila Valieva consent to using trimetazidine (see Introduction)?

The Bay Area Laboratory Co-operative (BALCO) scandal that broke in 2003 contributed to yet another shift in anti-doping ef-forts. In brief, investigators uncovered that BALCO, a small California nutritional supplement company, had been providing some of the world's top athletes with performance-enhancing drugs, including the anabolic steroid tetrahydrogestrinone (THG), colloquially referred to as "The Clear." What made this case so remarkable was that rather than using an existing drug developed for medical use, THG was the first known designer drug developed specifically for sport. At that time, there was no way for WADA to test for a drug without first studying its composition. The test for THG could only be developed after a spurned track coach sent a sample to the United States Anti-Doping Agency.[57]

The traditional model for drug testing was reactive—it reacted to the presence of a known substance. The BALCO affair underscored the need for proactive testing, which manifested in the Athlete Biological Passport (ABP), first implemented by WADA in 2009. Rather than testing for the presence or metabolites of a substance or method in an athlete's blood or urine, the ABP tracks a longitudinal record of an athlete's biochemistry. Suspicious fluctuations can suggest doping, so rather than focusing on identifying a substance in an athlete's blood and urine, testers look for the effects of that substance on the body's chemistry. It is a complicated and ongoing project that has shown some promise, but WADA's primary mechanism remains testing for the substances and methods it places on its annually updated Prohibited List.[58]

What to Prohibit and Why?

WADA's 2024 Prohibited List consists of six categories of substances and three varieties of methods that are always banned, both in and out of competition, including anabolic steroids and gene doping (Chapter 5). Another four categories of substances are prohibited only in competition, such as stimulants and narcotics, while beta blockers, which lower a person's heart rate or blood pressure, are only banned in certain sports where too much adrenaline is a disadvantage, such as shooting and archery, which require a steady, even hand (Table 4.2).[59]

According to the Code, a substance or method "shall be considered for inclusion on the Prohibited List if WADA, in its sole discretion,

Table 4.2 WADA's Prohibited List

Prohibited Substances (at all times; in and out of competition)

S0. Non-Approved Substances

S1. Anabolic Agents

- Anabolic Androgenic Steroids
- Other anabolic agents

S2. Peptide Hormones, Growth Factors, Related Substances, and Mimetics

- Erythropoietins (EPO) and agents affecting erythropoiesis
- Peptide hormones and their releasing factors
- Growth factors and growth factor modulators

S3. Beta-2 Agonists

S4. Hormone and Metabolic Modulators

- Aromatase inhibitors
- Anti-estrogenic substances (anti-estrogens and selective estrogen receptor modulators)
- Agents Preventing Activin Receptor IIB Activation
- Metabolic Modulators

S5. Diuretics and Masking Agents

Prohibited Methods (at all times; in and out of competition)

M1. Manipulation of Blood and Blood Components

M2. Chemical and Physical Manipulation

M3. Gene and Cell Doping

Prohibited Substances (in-competition only)

S6. Stimulants

S7. Narcotics

S8. Cannabinoids

S9. Glucocorticoids

Prohibited in Particular Sports

P1. Beta-Blockers: Archery★, Automobile, Billiards, Darts, Golf, Shooting★, Skiing/ Snowboarding. Underwater sports

★Prohibited in-competition and out-of-competition (no asterisk indicates the substance is only prohibited in-competition)

Adapted from The World Anti-Doping Agency, Prohibited List, 2024, https://www.wada-ama.org/sites/default/files/2023-09/2024list_en_final_22_september_2023.pdf.

determines that the substance or method meets any two of the following three criteria":

- Medical or other scientific evidence, pharmacological effect or experience that the substance or method, alone or in combination with other substances or methods, has the potential to enhance or enhances sport performance;
- Medical or other scientific evidence, pharmacological effect or experience that the Use of the substance or method represents an actual or potential health risk to the Athlete;
- WADA's determination that the Use of the substance or method violates the spirit of sport described in the introduction to the Code.[60]

Put simply, officials ask if a substance or method enhances performance, if it risks athletes' health, and if it violates the "spirit of sport." Answering yes to two out of the three questions is grounds for prohibition, but each question has been the subject of justifiable scrutiny.

Performance Enhancement

The first criterion WADA officials consider is whether a substance or method "has the potential to enhance or enhances sport performance."[61] Bioethicists Silvia Camporesi and Mike McNamee take issue with the word "potential," noting that WADA "does not require that a substance have a demonstrably performance-enhancing effect for it to be included on the Prohibited List." A presumption of performance enhancement suffices. In fact, a review of published literature found "no convincing evidence for performance enhancement" in most of the prohibited substances.[62] This is not the same as saying that the drugs do not work, but rather that we lack proof, primarily because medical ethics prevent controlled, rigorous research studies on doping in elite athletes. Instead, policymakers make presumptions based on clinical trials conducted on nonathletes for therapeutic purposes, anecdotal evidence, and cautionary, often mythologized tales.[63]

Leaving aside the lack of scientific evidence, one might suppose that nearly everything an elite athlete does is designed to enhance performance. As political scientist David van Mill posits:

> Athletes try to enhance their performance in many ways: coaches, psychologists, dietitians, massage therapists, training at high altitude, skin-tight swimsuits. All of these are used to gain an advantage, which is often unfair because, like drugs, they are available to some—wealthy athletes rather than cheats—but not to everyone.[64]

When it comes to anti-doping, enhancement means something different from improvement or optimization. Taken in concert with the "spirit of sport," enhancement exceeds the "perfection" of talent within one's "natural" limits.[65] Philosopher Torbjörn Tännsjö defines enhancement as that which "aims at taking an individual beyond the normal functioning of a human organism."[66] Enhancement, then, is constructed against the nebulous concepts of "natural" and "normal." Enhancement is unnatural. It abnormally augments performance. Critics insist that allowing enhancement will either alter the "future of human nature" or lead to the "dehumanization" of sport and society.[67]

One could argue that no body is natural, at least in postmodern societies. Consider, as examples, the ubiquity of immunizations, medical interventions, or the chemicals we ingest in our foods. In terms of athletic performance, maintains one research team, "sports have never been a test of merely 'natural' capabilities, but that they have always been constitutively technological, whether this involves specific artefacts or simply the application of scientific knowledge."[68] Ethicist Thomas Douglas puts it simply: "There is no sport in which natural ability is the only determinant of success."[69] And it is worth noting that sport's insistence on natural bodies contradicts World Athletics' regulations that require women with certain DSDs to suppress their endogenous testosterone (see Chapter 3).

One could similarly argue that elite athletes are not normal. As Dr. J. G. P. Williams wrote in his 1962 textbook *Sports Medicine*, the top athlete is "as different physiologically and psychologically from the 'man on the street' as is the chronic invalid."[70] Those differences have only widened since Williams's observation. As sport has gained popularity, prestige, and profitability, athletes' bodies have become increasingly specialized and distinctly abnormal.

What counts as enhancement, then, is about surpassing the conditions specified in protective policies. Those conditions can seem somewhat arbitrary. Athletes with unimpaired vision commonly turn to LASIK surgery and specialized contact lenses to make their eyesight "better than perfect."[71] These are vision enhancement techniques, as opposed to therapeutic or corrective measures, and can confer a disputably unnatural advantage. Mark McGwire, who in 1998 broke Major League Baseball's single-season homerun record, wore contact lenses that not only corrected his impaired vision but took it beyond "normal" functioning to 20/10. He could see at twenty feet what the average person could see at ten, an obvious advantage for someone trying to hit a baseball screaming toward the plate.[72] For all the outcry over McGwire's use of performance-enhancing drugs, no one objected to his use of performance-enhancing contacts.

When it comes to our tolerance for enhancement, we hold sport to a different standard than other areas of social life. Scholars posit that we are less accepting of performance enhancement in sport than in areas such as personal appearance, the arts, or sex because of what Tännsjö identifies as the "ethos of elite sport," which values aesthetics, competitiveness, and fairness.[73] Yet sport is fundamentally unfair. The playing field is irredeemably unlevel, and most inequities go unregulated.

Professional basketball players are remarkably tall; Olympic gymnasts are unusually short; elite sprinters tend to have at least one copy of the R-variant ACTN3 gene; five-time Tour de France winner Miguel Indurain reportedly has "Zeppelin-sized lungs, piston-like femurs" and a "resting heart rate of just 28 beats per minute (the adult norm is between 60 and 90bpm)."[74] Athletes with these attributes are granted advantages over those without them.

But social differences, not biological differences, result in the greatest athletic inequities. Who makes it to the Olympic Games, for instance, is influenced by race, ethnicity, country of origin, and socio-economic status. According to a 2017 study, both summer and winter Olympic sports "predominantly" favor "white and privately educated Olympic athletes."[75] The same is true for Paralympians, where the use of sport-specific technologies, such as wheelchairs, bikes, and prostheses, puts the price of competition well out of reach of the majority of people with impairments.[76] Even WADA's Ethics Panel acknowledges that "individuals are fundamentally different and not equal, and the circumstances under which athletes might have to train are not the same (e.g. due to differences in resources)."[77] Yet there are no policies in place to protect athletes against social inequities, which presumably strike the leaders of elite sport as normal and natural.

Health

The second criterion WADA considers when assembling its prohibited list is whether a method or substance "represents an actual or potential health risk to the *Athlete*."[78] Again, the term "potential" suggests the absence of conclusive evidence. There is evidence, however, that substances banned by WADA have been used to great therapeutic effect. For this reason, WADA allows athletes with a medical need to apply for a Therapeutic Use Exemption so that they can take an otherwise banned substance to "return to the athlete's normal state of health."[79] Athletes with asthma, for example, have been approved to take salbutamol, which WADA bans as both a stimulant and an anabolic agent. Athletes with diagnosed growth deficiencies, such as soccer great Lionel Messi, are permitted to take human growth hormones. While there are instances in which athletes have abused the system, there are other instances when athletes have been denied an exemption, forcing them to choose between competing in sport and medically necessary treatment.[80] This possibility increases for Para athletes, who may have health conditions that require special considerations.

Further complicating the "health" criterion is that anti-doping practices may "paradoxically introduce more health problems than they prevent," writes one research group.[81] These problems include doping without adequate understanding of the drugs and processes, the turn to black-market and more dangerous drugs and methods, sharing needles with other athletes, the use of unregulated masking agents, and more.[82]

An even bigger paradox, argue critics, is that elite sport is inherently unhealthy—a deep if rarely acknowledged problem. Top athletes appear to be the picture of good health: they are lean, strong, powerful, flexible, vital, and resilient; biomechanically efficient, physiologically fine-tuned, nutritionally optimized, and psychologically honed. Even so, scientists distinguish between "health"—a state of mental, social, and physical well-being—and "fitness"—the quality of being able to perform a specific physical task. "Too many athletes," they conclude, "are fit but unhealthy."[83]

More stridently, sociologist Kevin Young asserts that elite sport "is a violent and hazardous workplace, replete with its own unique forms of 'industrial disease.'"[84] Whether training "constantly at or near a state of physical breakdown," playing through injury, pathological eating behaviors, the use and abuse of supplements, painkillers, and anti-inflammatories, or the chance of living with chronic and debilitating pain, being a top athlete "isn't actually a super-healthy thing to do," observes Olympic cyclist Mara Abbot.[85] Thus, as bioethicist Norman Frost concludes, using health as a rationale against doping is "disingenuous and misplaced for the risks of sport itself far exceed the demonstrated risks of those drugs that arouse the greatest concern."[86]

The Spirit of Sport

Of the three criteria WADA considers in banning a substance or method, the "spirit of sport" is the most contested.[87] The phrase does a lot of heavy lifting, not only as a criterion used to judge substances and methods, but as WADA's "fundamental rationale" against performance enhancement. As defined in the Code, the spirit of sport is "the ethical pursuit of human excellence through the dedicated perfection of each Athlete's natural talents." It is "the essence of Olympism and is reflected in the values we find in and through sport." Those values include:

- Health
- Ethics, fair play and honesty
- Athletes' rights as set forth in the Code
- Excellence in performance
- Character and Education
- Fun and joy
- Teamwork
- Dedication and commitment
- Respect for rules and laws
- Respect for self and other Participants
- Courage
- Community and solidarity[88]

The spirit of sport offers an aspirational, perhaps utopian vision—a normative "ideal," McNamee explains. The "values and virtues listed characterize sport at its best: this is what we ought to aim for."[89] According to WADA, doping corrupts this ideal.

Philosopher Claudio Tamburrini disagrees. Doping "is not only compatible with, but also incarnates, the true spirit of modern competitive elite sports." The use of banned substances and techniques, he contends, is "obviously in accordance with the 'spirit' of today's crudely competitive and highly technified sports world, as they have everything to do with the essential purpose of athletic contest: to expand the limits of our capabilities."[90] If elite sport is really about expanding our limits, as Tamburrini argues, then protective policies stand in the way.

The debate over the use of hypoxic (low oxygen) devices and environments illustrates the complexities of the "spirit of sport." These technologies mimic the effects of the "live high-train low" approach to improving performance: "Living high," that is, living at high altitude where the oxygen is low, triggers the body's production of red blood cells; "training low" allows athletes to work out in less physiologically taxing environments so they can maintain or increase the intensity of their workouts.

In the early 1990s, vexed by Finland's low elevation and eager to capitalize on the live high-train low formula, exercise scientist Heikki Rusko diluted the oxygen concentration in a room to create a hypoxic environment. Endurance athletes lived and slept for fourteen to eighteen hours a day in the "altitude house" and performed their normal training at sea level. It worked. The athletes increased their red blood cell mass and improved performance.[91] Before long, scientists had developed hypoxic tents, chambers, and masks to achieve similar results.[92]

In 2006, WADA officials took on the topic of artificially induced hypoxic conditions and divided the deliberation of its three criteria— performance enhancement, health, and the spirit of sport—between subgroups. Scientific committees took on the first two principles and resolved that the hypoxic devices potentially enhance performance but do not pose a health risk to users.[93] It was therefore up to WADA's Ethics Issues Review Panel to ascertain whether the technology violates the spirit of sport, and therefore meets the two-out-of-three rule that would prohibit its use.

The members of the ethics panel "concluded unanimously" that hypoxic devices contravened the spirit of sport. The crux of their decision rested on the distinction between "active" and "passive" technologies. Specifically, the experts reasoned that the "athlete is merely a passive recipient" of artificially induced hypoxic conditions, as opposed to "technologies with which the athlete actively engages and interacts as part of the process of training and competing."[94] Hypoxic devices, the group surmised, require no more effort than "entering a room, donning a mask, and flipping a switch."[95]

The panel's decision was steeped in "moral language," notes bio-ethicist Andy Miah.[96] As the members of the Ethics Issues Review Panel explained:

> The spirit of sport, as we understand it, celebrates natural talents and their virtuous perfection. We say "virtuous" in this context because virtues are qualities of character admirable in themselves, the qualities that outstanding athletes develop and embody in their quest for excellent performance.[97]

The use of hypoxic devices lacks virtue—a key if murky ingredient in WADA's recipe for the spirit of sport.

A global cast of critics quickly pointed out the flaws in the panel's logic. Approved training tactics, such as artificially cooling and heating the body, the use of saunas, massage, electrical stimulation, and even specialized diets and legal supplements are similarly "passive" but inspire no debate. Athletes can choose where they live and drive to sea level. Isn't "flipping a switch" on a hypoxic device the same thing as starting a car? In an editorial endorsed by seventy-six scientists from twenty-four countries, physician Benjamin Levine declared that "the passivity argument is biologically naïve, logically inconsistent, and scientifically untenable."[98] The same could be said for the spirit of sport.

After months of debate, WADA opted not to prohibit athletes from using artificially induced hypoxic technology. Still, qualified chairperson Richard Pound, "It doesn't mean we approve it."[99]

Possible Alternatives

In light of all the problems associated with anti-doping efforts—the cost, the effort, the unreliability, the many ways to beat the system, the porous line between therapy and enhancement, sport's ambiguous "spirit"—critics have wondered if the fight is worth it. Why not just adopt a position of "pharmaceutical libertarianism" and allow athletes autonomy and the right to self-determination, so long as their choices are educated and voluntary?[100] "Were we to treat athletes as mature adults capable of making informed decisions based on scientific in-formation," reasons sociologist Ellis Cashmore, "we could permit the use of performance enhancing substances, monitor the results and make the whole process transparent."[101] After all, doping alone will not make someone successful in elite sport. That person still must have talent, opportunity, and motivation. She must train hard and per-severe. Doping might just augment what already exists. Pragmatically, adds philosopher Thomas Douglas, "removing the anti-doping rules would completely remove the problem of cheating."[102]

Those who oppose the legalization of performance enhancement postulate that it would ruin sport as we know it, resulting in little more than a pharmaceutical arms race that would only exacerbate social inequities in favor of resource-rich athletes and nations.[103] In addition, athletes who object to doping will feel pressured to do so to remain competitive. The "coercion argument" is especially per-suasive in consideration of subelite and recreational athletes, as well as children, and other "vulnerable athletic and non-athletic popula-tions."[104] The cultural valuation of elite athletes and their physiques may prompt others to emulate their enhancement techniques.

Anti-doping proponents therefore call for "even more rigorous testing protocols," including "greater frequency of random doping analyses, enforced medical follow-ups, stronger legislation against the possession of doping substances, and harsher penalties for athletes who use the substances."[105] Of course, all of this would require added funds. WADA's 2022 reported budget was $46 million (USD). While that seems like an awful lot of money, experts contend that it is a mere fraction of the revenue generated by elite sport.[106] The IOC, which provides half of WADA's budget, generated $4.2 billion in revenue in 2021.[107] Presumably doping scandals erode the IOC's profit, and the Olympic movement has a vested interest in presenting an image of drug-free sport, despite evidence to the contrary.[108]

What is more, studies show that increased spending does not make anti-doping more effective.[109] All the money in the world won't help without committed support from International Federations, national governing bodies, national anti-doping agencies, and other stake-holders. In just one of too many incidents of socially organized dop-ing and institutional coverup, a 2015 investigation determined that the UCI colluded with Lance Armstrong and his lawyers to sup-press positive doping results. According to the report of the Cycling Independent Reform Commission:

> UCI saw Lance Armstrong as the perfect choice to lead the sport's re-naissance after the Festina scandal: the fact that he was American opened up a new continent for the sport, he had beaten cancer and the media quickly made him a global star. Numerous examples have been iden-tified showing that UCI leadership "defended" or "protected" Lance Armstrong and took decisions because they were favorable to him.[110]

Since stripped of his seven Tour de France victories, Armstrong was clearly not acting alone. Rather, his doping was facilitated by the very organization that was supposed to stop him. More money would not have prevented that from happening.

Finally, there are at least two alternatives between the pro-doping and anti-doping positions. The first is to allow athletes to dope under the supervision of physicians and "within the framework of classical medical ethical standards."[111] This would not only make doping safer, but it would also allow scientists to research the effects of substances and methods, and offer insights into the limits of human potential.[112] Medically supervised doping, contend scholars Eric Moore and Jo Morrison, "requires the acceptance that doping is not intrinsically bad, and that it is, in fact, at the very least consistent with some of the projects of competitive sport."[113]

Within this logic, supporters advocate for "harm-reduction" strat-egies that have had success within the broader field of public health, particularly with illicit drug use. To reduce the potential harms asso-ciated with doping, scholars Bengt Kayser and Jan Tolleneer propose a framework that includes "(1) the antidoping rule is relaxed within boundaries of acceptable health risks; (2) the athlete's health is moni-tored and (3) some urine and blood testing subsists using pragmatic evidence-based cut-off levels to control risk."[114] Bioethicist Julian

Savulescu offers a related position by advocating for "physiological doping," which he defines as "setting safe limits for physiological values such as testosterone levels," and allowing athletes to dope within those limits. "Testing then focuses not on how those levels were achieved, but on whether they are safe."[115]

McNamee calls Savulescu's position "naïve." Physiological doping, counters McNamee, "would not lead to a cleaner sport, but would lead to a two-tiered doping system, in which, in order to gain a competitive advantage, athletes would still take performance-enhancing substances covertly so as not to allow competitors to match their preparation."[116] In other words, efforts to make doping open and safe would have the opposite effect of making it even more clandestine and potentially dangerous.

A second alternative to the pro-/anti-doping debate is to create what sport historians Jan and Terry Todd call "parallel federations—one for drug users and another for those who choose to prepare for competitions without the use of ergogenic drugs." This is a shallow, if not superficial solution that does nothing to change the problems and unintended consequences of testing. Additionally, the Todds observe that parallel federations in powerlifting "failed because of an entirely unforeseen consequence." There was "no way to stop the proliferation of new federations once the idea of multiple federations took root." In 1981, frustrated with the US Powerlifting Federation's lack of testing, secessionists formed the American Drug Free Powerlifting Association. This "fractured" the sport, and by 2008 there were at least seventeen international and eighteen American powerlifting organizations, each with different rules, records, and standards.[117]

Nonetheless, parallel federations might reveal the true spirit of sport—or at least the spirit to which spectators and sponsors will gravitate. Commercial interest not only drives the market; it also plays a part in inducing athletes to dope. The demand for longer seasons, more arduous contests, the insistence that athletes play through pain and injury and extend their careers, and the call for athletes to continually perform swifter, higher, and stronger comes at a price.

5

Regulating Genetics

Eddy Curry's contract with the Chicago Bulls was set to expire at the end of the 2005 National Basketball Association (NBA) season. As the team's leading scorer, the twenty-two-year-old was in a good bargaining position, but an elite athlete's future is always fickle. During a game against the Memphis Grizzlies, Curry reported feeling chest pains. Doctors diagnosed him with an irregular heartbeat, and he sat out for the remainder of the season.

During Curry's subsequent contract negotiations, the Bulls' front office offered him a deal contingent upon genetic testing. Specifically, before issuing a new contract, management wanted to know if Curry was predisposed to hypertrophic cardiomyopathy, a heart condition responsible for the tragic deaths of too many young athletes, including the University of Loyola Marymount's Hank Gathers in 1990, and Reggie Lewis of the Boston Celtics in 1993, both of whom died on the basketball court.

Bulls general manager John Paxson insisted the team's concern was for Curry's protection. Curry countered that the test violated his privacy and that the results would affect his future in the league. Consenting would also set a dangerous precedent. "If employers could give employees DNA tests, then they could find out if there's a propensity for illnesses like cancer, heart disease or alcoholism," his lawyer, Alan Milstein, argued. "They will make personnel decisions based on DNA testing."[1] Curry refused the test, and the Bulls traded him to the

New York Knicks. He continued playing in the NBA until 2012, and then in Asia until 2019, without experiencing a major cardiac episode.

As the Curry-Bulls controversy played out, so did the tensions between Curry's autonomy and the Bulls' professed protectionism; between concern for his well-being and the team's desire to protect its best interests. There were also concerns that genetic testing could open the door to genetic discrimination. After all, as bioethicist Mark Rothstein reasons, "the most famous baseball player with a genetic disorder was Lou Gehrig. Would they have signed him if they knew he was predisposed to A.L.S. [amyotrophic lateral sclerosis, also known as Lou Gehrig's disease]?"[2]

These same tensions also characterize the National Collegiate Athletic Association's (NCAA) policy for testing athletes for sickle cell trait, established after the death of Dale Lloyd II. In 2005, the nineteen-year-old defensive back on Rice University's football team succumbed to complications from an undetected genetic mutation. Or perhaps he succumbed to complications from football's "toxic coaching culture."[3] Either way, the two complications collided on a hot and humid Sunday in Houston, Texas.

Lloyd and his teammates had completed their weight training and headed outside for a speed workout. There, coaches ordered the athletes to complete sixteen consecutive 100-yard sprints under the punishing September sun. Witnesses recounted that Lloyd quickly showed signs of distress. Usually one of the fastest players on the team, he lagged 30 to 40 yards behind the others. He had difficulty breathing, and then trouble standing, and then he could barely hold up his head. The coaching staff forbade the other players from helping him and insisted he finish the drill. When it was over, Lloyd collapsed and never regained consciousness. He died the next day.[4]

Doctors later determined that Lloyd had undiagnosed sickle cell *trait*, which is caused when an individual has one mutated copy of the hemoglobin-Beta (*HBB*) gene. Humans inherit two copies of most genes—one from their biological mothers and one from their biological fathers. Individuals who inherit two copies of the mutated

HBB gene have sickle cell *disease*, which is associated with anemia, stroke, chronic pain, organ damage, and early death. Unlike the disease, the trait doesn't usually produce any noticeable health concerns. However, under extreme conditions, including physical exertion, heat, and dehydration, the trait can cause "exertional sickling" in which the donut-shaped red blood cells collapse into the form of a crescent or sickle. The sickled cells build up in small blood vessels and decrease blood flow, leading to physical distress, collapse, and even death.

Upon learning of their son's diagnosis, Lloyd's parents filed a wrongful death lawsuit against Rice University, the coaching staff, the NCAA, and the makers of two nutritional supplements whose products contained Creatine, which may quicken dehydration. The family settled out of court for an undisclosed amount, but as their lawyer explained, it was "not about money; the suit is about policy."[5] As part of the NCAA's 2009 settlement with the Lloyd family, all incoming student-athletes are now screened for sickle cell trait.[6]

At first glance, screening for sickle cell seems to do far more good than harm. Researchers find that football deaths associated with the trait have decreased by 89 percent since the NCAA implemented its policy.[7] But there are important if unintended consequences to consider. To begin, those who test positive for sickle cell trait may unnecessarily avoid sport and exercise, while those who test negative may believe themselves impervious to hazardous athletic conditions.

The tests may also prove discriminatory. Coaches may not recruit or renew the scholarships of athletes with sickle cell trait, which is overrepresented in, but not exclusive to people of West African descent. As sociologist Troy Duster explains, "This could have an extraordinarily heavy impact on black athletes. You are going to be picking out these kids and saying, 'You are going to be scrutinized more closely than anyone else.' That's worrisome."[8]

Critics additionally worry that the protocol does not involve counseling to help athletes understand what their test results mean.[9] Testing, discussions, and decisions "should involve the athlete and his or her family and physician, not trainer or coach," advises Janis Abkowtiz, president of the American Society for Hematology. She and others claim that the NCCA's policy is "medically groundless—perhaps even dangerous—and is focused more on protecting the NCAA from legal liability than protecting the health of student athletes."[10]

Abkowtiz's larger point is that pathologizing a single gene does nothing to change a culture that habitually puts athletes in danger. It is a shallow solution to a deep problem. It matters that the only sickle cell-related deaths of American collegiate athletes have happened in football. It matters even more that all those deaths occurred during conditioning sessions. As the University of Oklahoma's head athletic trainer Scott Anderson points out, "We don't see exertional sickling in football games, ever ... If we're seeing it in our training and not seeing it in the sport, something is wrong with our training."[11] For these reasons, experts recommend that instead of genetic screening, the NCAA should "implement universal precautions to reduce exercise-induced injury." This would benefit all athletes, not just those with special genetic risks.[12]

Barring significant cultural change, another option might be to eradicate the harmful *HBB* gene altogether. In 2001, scientists at the Massachusetts Institute of Technology and Harvard University used gene therapy to successfully treat sickle cell disease in mice. The first human trials came eighteen years later, and dozens of patients have shown remarkable improvement. Researchers are reluctant to call them "cured," however. "We don't use the 'c word,'" explains stem-cell biologist Donald Kohn, "but they're looking really promising." It is too soon to tell if the patients will experience any long-term complications, and one company paused its clinical trials after two patients developed leukemia-like cancer.[13]

Such risks have not deterred athletes, coaches, and sports officials from pursuing the possibilities of "gene doping," which WADA first banned in 2003. As with doping more generally, gene doping applies therapeutic techniques to otherwise healthy athletes for the purpose of performance enhancement. While rumors of genetically modified athletes abound, to date, there have been no documented cases. It may just be a matter of time. As WADA chair Richard Pound quipped, "You would have to be blind not to see that the next generation of doping will be genetic."[14] If the blindness is genetic, there may soon be a cure.

This chapter addresses the regulation, and the lack of regulation, regarding three types of genetic testing: testing for athletic talent; testing for illness or injury; and testing for gene doping. Without getting too far into the complicated genetic science, the purpose is to mine these types of tests for legal, moral, ethical, and practical concerns. Chief among those concerns is the (mis)use of genetic information, the possibilities of genetic discrimination, questions of autonomy and consent, an athlete's right to privacy, and how and whether to enforce anti-gene doping policies.

Genetic Testing for Athletic Talent

With the exception of red blood cells, the nucleus of every cell in the human body contains chromosomes. One chromosome in every pair typically comes from each biological parent—twenty-three from the mom and twenty-three from the dad (see Chapter 3). Chromosomes are made up of genes, which are sequences of deoxyribonucleic acid (DNA). Genes provide the instructions for the production of proteins that perform various functions in the body. The *HBB* gene implicated in sickle cell trait, for example, tells the body to make a protein called beta-globin, a component of hemoglobin, the molecule that red blood cells use to bind and transport oxygen.

There are 20,000 to 25,000 genes in the human body, and we only know what about half of them do. This knowledge has accelerated rapidly since the 2003 completion of the Human Genome Project, which identified the genes in every chromosome and their chemical composition. That information has been invaluable as scientists seek to unlock the mysteries of the body—why we might look and behave certain ways, our propensities for disease and injury, how our bodies interact with specific medicines or nutrients, and what we might do to treat or even cure certain afflictions through genetic therapy and engineering.

The completion of the Human Genome Project has also brought new insights into athleticism. Within the relatively young field of "sport genomics," researchers estimate that there are more than 200 genes associated with physical performance and more than 20 genes connected to elite athletic status, which are sometimes referred to as "performance-enhancing polymorphisms" (genetic variations).[15] But we should be wary of the term, cautions Dr. Stephen Roth, an exercise physiologist and director of the Functional Genomics Laboratory at the University of Maryland. "The group of polymorphisms that have been associated with sport-related traits have such mixed findings that few would be comfortable calling them performance enhancing."[16]

There are certainly mixed findings when it comes to the *ACE* gene, which, in a 1998 study, was the first to be associated with athletic performance.[17] The *ACE* gene codes for the angiotensin-1 converting enzyme that helps regulate the circulatory system and blood pressure. There are two common variants: *ACE* I and *ACE* D. Humans get either an I or a D from their biological fathers and an I or D from their biological mothers, making three possible combinations for the *ACE* gene: two Is (I/I), two Ds (D/D), or one of each (I/D). The *ACE* I/I combination may help in endurance sports, while *ACE* D/D may be an asset in sports that emphasize strength and power.[18]

A similar pattern appears in the R and X variants of *ACTN3*, the gene that provides instructions for making the protein alpha-actinin-3, found almost exclusively in type II skeletal muscle fibers.[19] Often referred to as "fast-twitch" fibers, type II produce the quick, strong contractions needed for explosive speed, strength, and power in sprinting, jumping, and weightlifting. But they fatigue quicker and take longer to recover than type I or "slow twitch" fibers, which contract slowly, generate little power, and resist fatigue, making them valuable in endurance sports.

People have both types of fibers in their muscles, but the relative percentage of one type to the other matters. It also correlates with *ACTN3* polymorphisms: the R variant instructs the body to make the alpha-actinin-3 protein; the X variant prevents its production. Most elite strength and power athletes show at least one copy of the R variant—the so-called speed gene. Two copies may be better, while two copies of the X variant may prove advantageous in endurance sports.[20]

In 2004, just one year after the publication of the first "speed gene" study, the Australian company Genetic Technologies started selling its direct-to-consumer (DTC) "ACTN3 Sports Performance Test."[21] Consumers could order the kit, swab the inside of their cheeks, and send their specimens to a lab for analysis. Within a few weeks, Genetic Technologies would return a certificate announcing "Your Genetic Advantage" based strictly on its assessment of the *ACTN3* gene.

Critics quickly voiced their concerns. According to scholar Timothy Caulfield, "the relevance of the [*ACTN3*] gene—about 30 percent of the population carries the two copies of the variant associated with speed—to actual 'athletic ability' is greatly exaggerated by DTC companies." He continues:

> While this gene *is* related to the regulation of fast-twitch muscle fibers, it is wrong to imply that it is a test for athletic ability, a complex, socially constructed and multi-factorial concept, or that it can provide anything close to a definitive conclusion about future speed abilities. Athletes who do not have the alleged speed gene have made it to the Olympics in speed/power sports and many millions who do have the favored genetic allotment have languished in mediocrity.[22]

It is worth repeating: athletic ability is "a complex, socially constructed and multi-factorial concept" and one gene, on its own, will not make someone a good athlete.

Further, the "single-gene-as-magic-bullet" message used to sell DTC products distorts the interactions of different genes with each other.[23] Athleticism is not *monogenic*—that is, something that is linked to a single gene, as it is in sickle cell, but rather *polygenic*—something affected by myriad genes. "We've been realizing, and it's just been borne out over the past several years, that it's not on the order of 10 or 20 genes but rather hundreds of genes, each with really small variations and huge numbers of possible combinations of those many, many genes that can result in a predisposition for excellence," clarifies Roth.[24]

It is not only the combinations of genes that count but also the crucial interactions between those genes and the environment—an extraordinary mix of nature and nurture. "If you could equalize all environmental factors, then the person with some physical or mental edge would win the competition," says Roth. "Fortunately, those environmental factors *do* come into play, which gives sport the uncertainty and magic that spectators crave."[25] No matter how biologically predisposed one is to athletic greatness, that greatness also requires the interest, opportunity, means, and support to materialize.

For these and other reasons, some experts maintain that DTC genetic tests are nothing more than modern-day "snake oil," and that manufacturers misrepresent and manipulate scientific knowledge for commercial gain.[26] In response, Deon Venter, director of Genetic Technologies, claims, "It's not a test that says you are going to be a winner or a loser. It's a test that appears, on the evidence we have so far, to head people into choosing the best event and in some cases the optimal sport."[27]

Importantly, Venter calls it "the evidence we have so far." In the decades since the launch of the "ACTN3 Sports Performance Test," that evidence remains inconclusive. For the most part, this is true of all the genes associated with athletic performance. Findings are often contradictory, have little predictive value, and are derived from relatively small studies that lack statistical power.[28] These same limitations plague the broader DTC genetic testing industry—whether testing for ancestry, a genetic predisposition to disease, the potential for food allergies, and all points in between (see Table 5.1). For these and other reasons, France, Germany, Portugal, Switzerland, and several other countries only allow medical doctors to conduct genetic testing.

Table 5.1 Critiques of Direct-to-Consumer (DTC) Genetic Testing

The DTC genetic testing industry lacks regulatory and professional oversight.

There is not enough existing evidence for clinical validity and utility.

DTC companies misrepresent or distort the scientific evidence through "deceptive marketing and other questionable practices."*

Consumers may misinterpret their test results.

Relatively few DTC consumers share their results with genetic counselors or other healthcare providers to help them make sense of the analysis.

Test results are often inconsistent and inaccurate, resulting in both false positives and false negatives.

Consumers are not assured the right to privacy. Even if their results are de-identified, each individual genome is unique.

There are possibilities for genetic discrimination, particularly if a consumer's information is made available to employers, insurers, or other parties.

DTC companies often sell consumers' information to third parties, such as pharmaceutical and biotechnology companies, or are made available to law enforcement agencies. Consumers typically grant consent for this, but there are questions about how informed they are when they do.

DTC tests are not covered by insurance, and consumers foot the bill, which means that not everyone has equal access to testing.

See Loredana Covolo, Sara Rubinelli, Elisabetta Ceretti, and Umberto Gelatti, "Internet-Based Direct-to-Consumer Genetic Testing: A Systematic Review," *Journal of Medical Internet Research* 1, no. 12 (2015): e4378; Mary A. Majumder, Christi J. Guerrini, and Amy L. McGuire, "Direct-to-Consumer Genetic Testing: Value and Risk," *Annual Review of Medicine* 72 (2021): 151–166; Andelka M. Phillips, "Buying Your Genetic Self Online: Pitfalls and Potential Reforms in DNA Testing" (2015): 77–81; Paula Saukko, "State of Play in Direct-to-Consumer Genetic Testing for Lifestyle-Related Diseases: Market, Marketing Content, User Experiences and Regulation," *Proceedings of the Nutrition Society* 72, no. 1 (2013): 53–60; Amy B. Vashlishan Murray, Michael J. Carson, Corey A. Morris, and Jon Beckwith, "Illusions of Scientific Legitimacy: Misrepresented Science in the Direct-to-Consumer Genetic-Testing Marketplace," *Trends in Genetics* 26, no. 11 (2010): 459–461 Nicole Vlahovich et al., "Genetic Testing for Exercise Prescription and Injury Prevention: AIS-Athlome Consortium-FIMS Joint Statement," *BMC Genomics* 18, no. 8 (2017): 5–13; Timothy Caulfied, "Predictive or Preposterous? The Marketing of DTC Genetic Testing," *Journal of Science Communication* 10, no. 3 (2011): 1–6.

* Gregory Kutz, *Direct-to-Consumer Genetic Tests: Misleading Test Results Are Further Complicated by Deceptive Marketing and Other Questionable Practices: Congressional Testimony* (DIANE Publishing, 2010).

None of this has stopped the sport-related DTC industry from forging ahead and expanding its purview beyond the "speed gene." As of 2019, almost seventy companies marketed genetic tests for sport, exercise performance, or injury, with prices that ranged from $100 to $1,100.[29]

These types of tests are especially troubling when geared toward children. Atlas Sports Genetics, for example, claims that it "Gives parents and coaches early information on their child's genetic predisposition for success in team or individual speed/power or endurance sports."[30] One American couple took advantage of this "early information" by testing their thirteen-month-old daughter's *ACTN3* constitution. "If she came back all endurance, we'd probably focus more on the long-distance type things," the mother told journalists. "Likewise, if she was all strength, we would direct her toward power sports." It may be lucky for the toddler that the results came back "a mix," as the mother put it, suggesting a combination of the R and X variants.[31] Maybe she'll be able to "focus" and "direct" herself down paths of her own choosing.

Ethicists warn that based on the results of these tests, parents or coaches may guide or force children away from certain sports and into others, or even out of sport entirely. This could sacrifice the children's right to an open future and engagement in activities that could bring lifelong satisfaction. And there are serious questions about whether children can consent to the tests and to what degree the adults in that child's life understand the limited and ambiguous science behind them.[32]

Proponents of the tests reason that many parents already engage in different forms of talent identification and navigate their kids accordingly. "Children with nimble fingers and perfect pitch are encouraged to play the violin and children who grow tall at a young age are encouraged to play basketball," argue bioethicists Julian Savulescu and Bennett Foddy.[33]

Perhaps, but several groups, including the Australian Institute of Sport and the American Society of Human Genetics, recommend that the predictive talent tests are not appropriate for individuals under the age of eighteen.[34] As a venerable team of scholars asserted in a 2015 consensus statement:

> Genetic tests have no role to play in talent identification or the individualized prescription of training to maximize performance. . . . Consequently, in the current state of knowledge, no child or young athlete should be exposed to genetic testing to define or alter training or for talent identification aimed at selecting gifted children or adolescents.[35]

WADA is less assertive, stating only that "the use of genetic information to select for or discriminate against athletes should be strongly discouraged."[36] This does little to deter the practice; it's even been activated on a national scale. In 2014, Uzbekistani officials announced a plan to test children as young as ten years old in an effort to boost the country's performance in international sport. The children's "parents will be told what sports they are best suited for."[37] In 2018, China's Ministry of Science and Technology similarly announced that athletes hoping to compete in the 2022 Winter Olympics would be selected after undergoing "complete genome sequencing" to test for "speed, endurance and explosive force."[38] There was no report of how the procedure affected results, although China did finish third in the medal count. While these methods of talent identification and development may produce the desirable results, what might be sacrificed in the process?

Testing for Injury and Illness

There are certainly less controversial uses for genetic information in sport. Bioethicists, sports scientists, and administrators are generally more amenable to tests relating to an athlete's propensity for illness or injury, a growing practice in the broader field of medicine known as "risk prediction." In sport, the use of these tests could assist in personalizing athletes' training, transferring them to a sport or event where they might be more successful, prescribing an optimal diet, or avoiding certain injuries. It's already happening in everything from the Australian Rugby League, to the English Premier League soccer, to Baylor University football. British long jumper Greg Rutherford used his DNAFit results to alter his training in advance of the 2016 Olympic Games. "The DNA showed I have a level of endurance in me which I never really realized," he told journalists. He changed his routine to include "things like running hills and larger amounts of reps in the gym" and went on to earn a medal at the 2016 Games in Rio.[39]

Knowledge about athletes' genetic makeup will not prevent them from getting hurt, but it may help them decrease the likelihood of sustaining particular injuries through adapted training, extra precautions, and prehabilitation. For example, a 2018 study associated several polymorphisms with an increased risk of stress fractures. Athletes showing those polymorphisms could take preventative steps, such as additional bone mineral density monitoring, and vitamin D and calcium supplementation.[40] Other genes are involved with the production of collagen, the main component of tendons, ligaments, bones, and other tissues. Variations in specific genes have been linked to a greater risk for anterior cruciate ligament rupture, tendinopathy, and shoulder dislocations—potentially season- and even career-ending injuries.[41] When NFL offensive lineman Andy Alleman learned he had a version of the *MMP3* gene that compromised his Achilles' tendon, he added exercises to increase its flexibility.[42] It's hard to argue against that line of defense.

Within this research, the "global concussion crisis" is a growing area of interest. Because some of the symptoms of chronic traumatic encephalopathy (CTE) look a lot like Alzheimer's disease, researchers speculate that the two conditions might share a genetic relationship. Specifically, the *APOE* e4 variant associated with Alzheimer's might have something to do with who develops CTE. The working hypothesis is that athletes with the variant may have a worse reaction to a traumatic brain injury and more difficulty recovering from it, although scientists warn that "we should be cautious in considering them genetic risk factors" until more conclusive studies emerge.[43]

Even so, England's Rugby Football Union reportedly planned to test all professional players for *APOE* e4. The players rejected the proposal "out of hand," one insider disclosed. They "had serious reservations about giving up sensitive personal information which could have been used against them in contract negotiations or by unscrupulous insurance companies. We were promised the study would be anonymous but no one bought that. It just felt wrong on a number of levels."[44] Similar to the Curry-Bulls controversy and the NCAA's test for sickle cell trait, there are critical questions about who owns the results of those tests and what can be done with them.

Also similar to the NBA and NCAA are questions about what motivates genetic testing in the first place—is it to protect the best interest of the players or the best interest of the organization? In 2015 alone, leagues spent hundreds of millions of dollars on the salaries of sidelined players: $300 million in the English Premier League; $350 million in the NBA; over $450 million in the NFL; and more than $700 million in Major League Baseball.[45] Armed with knowledge about a player's genetic makeup, some of those injuries and the associated costs might be avoidable.

While it is possible that Rugby Football Union officials proposed the *APOE* e4 tests out of genuine concern for the players' health, it also happened under the looming threat of a lawsuit. Veteran players have intimated their intent to sue the union, as well as World Rugby, Rugby Football League, and the Welsh Rugby Union for failing to protect them from the health risks caused by traumatic brain injury.[46] Precedent suggests they may have a case. Just a few years earlier, the NFL agreed to a $1 billion settlement with former players for a similar claim. Arguably, then, genetic screening might do more to protect the organizations' pockets than the athletes who line them.

As our genetic knowledge advances, so, too, does the field of gene therapy—the modification of genes to treat illness or disease. Conceptualized in the 1970s, the first approved gene therapy trial began in 1990. Since then, advancements in the field have been a boon for people suffering from a variety of ailments, but they have also sounded alarms about genetic engineering, modern eugenics, and "genetically modified athletes," to use bioethicist Andy Miah's term.[47] While it sounds like the stuff of science fiction, we may be closer to that reality than we realize.

Genetically Modified Athletes

Genetic manipulation existed centuries before scientists began to unlock the riddles of the human genome. It was just called breeding.
Breeders experimented with bringing together different plants or
animals to produce desired characteristics. In the mid-1800s, for example, livestock owners developed a line of Belgian Blue cattle that
were "double-muscled" to produce lean and tender meat. To look
at the creatures, the term "double-muscled" seems accurate, but it's
not. Belgian Blues have the same number of muscles as conventional
cattle, but those muscles develop more and bigger muscle fibers (conditions called hyperplasia and hypertrophy, respectively), which causes
them to swell to enormous proportions. It would take 150 years for
scientists to associate this trait with the *MSTN* gene and to consider
its possibilities for athletic performance.

In 1997, a research team at Johns Hopkins University discovered
the Belgian Blues' genetic secret while deleting specific genes in mice.
The process produced "mighty mice," with leg and chest muscles that
were up to four times larger than those of the control mice. Deleting
the previously unidentified gene silenced the production of a skeletal muscle-regulating protein the researchers called myostatin (*myo*
means muscle in Latin; *statin* means to halt). Without the protein that
told the muscles to stop growing, the "myostatin knock-out mice" developed more and larger muscle fibers and were otherwise healthy.[48]

The subsequently named *MSTN* gene also affects the racing performance of whippets. In trying to produce the fastest racing dogs, breeders
sometimes got excessively muscular "bully whippets." The bullies show
no significant health concerns, but they make lousy racers—they are
too stocky and tend to cramp in the shoulders and thighs. The authors
of a 2007 study determined that bully whippets have two copies of an
MSTN mutation, but dogs that carried just one copy of the mutation
were among the fastest at the highest levels of racing.[49]

As *MSTN*-related studies on animals developed, an exceedingly muscular baby caught the attention of Markus Schuelke, a German pediatric neurologist. Upon sequencing the boy's DNA, Schuelke found that both copies of his *MSTN* gene were inactive. The boy produced no myostatin.[50] Schuelke kept the boy's identity confidential, and it remains to be seen what type of athletic future he might have, but the exercise, fitness, and sports industries are now flush with substances and methods that claim to suppress myostatin.

As summarized in Table 5.2, there are a number of single-gene candidates associated with the possibility of doping. This includes the *IGF-1* gene, which encodes for the protein insulin-like growth factor 1 that boosts muscle growth and helps their repair. In the late 1990s, molecular physiologist Lee Sweeney and colleagues used a common virus to insert a synthetic *IGF-1* gene into the DNA of the muscle cells of mice. The process triggered the extra production of the growth factor to create "Schwarzenegger mice." Younger mice increased in muscle mass and strength by about 15 percent and older mice regained lost strength to become approximately 30 percent stronger than they were before the treatment.[51] Sweeney's results, as with previous research into the *MSTN* gene, have enormous implications for treating muscle-wasting diseases like muscular dystrophy, helping people to recover from an injury, and preventing muscle degeneration in aging patients. They may also have enormous implications for sport.

Not long after Sweeney published his findings, the sports world came calling. During an Olympic year, he receives three to four contacts a day from athletes asking him to enhance their genetic profile. One caller offered him $100,000 for his services. A junior college football coach approached Sweeney about the possibility of genetically altering his entire team.[52] Sweeney has been so deluged by these requests that he developed a stock response: "I basically say this is experimental. It's in animals, and even if I had it available to give to humans, it has to go through clinical trials to make sure it's safe."[53]

Yet safety often takes a backseat to athletic ambition. Riding shotgun might be experiments with the *EPOR* gene, which provides

Table 5.2 Potential Doping Genes

Abbreviation	Name	Expected Physiological Response	Potential for Performance Enhancement
EPO	Erythropoietin	Stimulates red blood cell production; increases blood oxygenation	Increased endurance
HIF-1	Hypoxia-inducible factor 1	Regulates transcription at hypoxia response elements (EPO, VEGF); increased number of red blood cells; increased blood oxygenation	Increased endurance
IGF-1/GH	Insulin-like growth factor 1/Growth Hormone	Regulates cell growth and development (IGF-I); increases lipolysis, protein synthesis, and glycogenolysis; growth of bones and tissue mass	Increased muscle power and mass
MSTN	Myostatin	Negatively regulates muscle cell grow	Increased muscle mass and strength
VEGF	Vascular endothelial growth factor	Stimulates angiogenesis and vasculogenesis	Increased endurance
PPAR δ	Peroxisome proliferator-activated receptor	Regulates the oxidation of fatty acids and increases mitochondrial activity and muscular glucose uptake; enhances slow-twitch muscle fibers and decreases fast-twitch fibers	Increased speed and endurance
ACE	Angiotensin-converting enzyme	Regulates blood pressure by adjusting angiotensin II levels and increases the proportion of slow-twitch muscle fibers	Increased endurance
POMC/PENK	Endorphin/ enkephalins	Reduces pain and fatigue threshold	Increased endurance
PEPCK-C	Phosphoenolpyruvate carboxykinase	Regulates gluconeogenesis and is involved in Krebs cycle	Increased endurance
ACTN2 and ACTN3		Increased rate of glucose metabolism in response to training (ACTN3); compensation for loss of function of ACNT3 gene by ACTN2 gene	Increased endurance, muscle strength and speed of muscle; increased efficiency in sprinters

Adapted from Olivier Salamin, Tiia Kuuranne, Martial Saugy, and Nicolas Leuenberger, "Loop-Mediated Isothermal Amplification (LAMP) as an Alternative to PCR: A Rapid On-site Detection of Gene Doping," *Drug Testing and Analysis* 9, no. 11–12 (2017): 1731–1737; Ewa Brzezianska, Daria Domanska, and Anna Jegier, "Gene Doping in Sport—Perspectives and Risks," *Biology of Sport* 31, no. 4 (2014): 251.

instructions for making a protein called erythropoietin receptor. As described in Chapter 4, erythropoietin (EPO) is the hormone that signals the body to produce red blood cells. To trigger that production process, erythropoietin first attaches to the erythropoietin receptor. Certain *EPOR* variants can result in polycythemia, an increased concentration of hemoglobin, the oxygen-carrying protein in red blood cells, which can foster endurance. It is why athletes blood dope.

Nordic ski champion Eero Mäntyranta didn't need to dope. His body naturally produced as much as 65 percent more red blood cells than the average man due to a genetic condition called primary familial and congenital polycythemia.[54] Although the Finnish wonder denied this contributed to his success, author David Epstein christened it a "gold medal mutation."[55]

It is a mutation that scientists hope to copy to treat anemia. In the late 1990s, scientists injected baboons with a genetic compound to boost EPO production. It worked. The animals' red blood cell counts nearly doubled within ten weeks. But it caused the blood to be so thick that the scientists had to regularly dilute it to keep the primates alive.[56] In 2004, scientists were likewise able to produce "supraphysiologic levels of EPO and polycythemia" in genetically altered monkeys.[57] Yet ironically, some of the animals developed severe anemia. Their immune systems wiped out not only the EPO introduced by gene therapy but their natural EPO as well. These outcomes continue to pose serious challenges to would-be gene dopers.

Despite these and other cautionary tales, the history of doping has shown that athletes are willing to risk their health in the pursuit of better performance. Indeed, there have been rumors of genetically modified athletes since the 2000 Summer Olympic Games in Sydney. The rumors intensified during the 2006 trial of German athletics coach Thomas Springstein, accused of doping young runners without their consent ("He said they were vitamins," one athlete told reporters, recalling the earlier East German program). During their investigation, police uncovered an email from Springstein reading, "The new Repoxygen is hard to get. Please give me new instructions soon so that I can order the product before Christmas."[58]

Raids of Springstein's home did not find any Repoxygen, a prototype gene-therapy drug developed to treat anemia by boosting red blood cell production. Nevertheless, athletes may be experimenting with it or something similar. After all, the formula for Repoxygen is publicly accessible and several websites advertise its sale. Then again, WADA's science director Olivier Rabin bought a few samples from these sites. "What came was just versions of synthetic EPO," he reported.[59] Doper beware.

To address the eventuality of gene doping, the IOC's Medical Commission first convened a "Gene Therapy Working Group" in 2001. The following year, WADA hosted a similar meeting, which resulted in the addition of "gene doping" as a "prohibited method" in the World Anti-Doping Code—just as scientists announced the completion of the Human Genome Project.[60] As the science marches on, WADA has scrambled to update the Code with the latest developments in gene therapy.

Most recently, this has involved WADA's prohibition of the "use of nucleic acids or nucleic acid analogues that may alter genome sequences and/or alter gene expression by any mechanism," many of which are associated with CRISPR-Cas9 technologies.[61] Briefly, CRISPR-Cas9 technologies utilize "molecular scissors" to cut a patient's DNA at desired locations and then repair these breaks to create desired "edits" (CRISPR stands for Clustered Regularly Interspaced Short Palindromic Repeats; Cas9 is a modified protein that snips the DNA's strands).[62] This allows scientists to change the DNA code in targeted ways to better understand and treat genetic diseases. It is easier and less expensive than other technologies and has had remarkable results in trials on microbes, animals, and as of 2016 in humans, including those with sickle cell disease.

At the forefront of this research is Dr. Matthew Porteus, a physician specializing in pediatric stem cell transplant and a pioneer in the field of genome editing and gene therapy. His work caught the attention of WADA, which enlisted him to serve on its Gene and Cell Doping Expert Group. Gene doping "will happen," Porteus predicts, but he qualifies that the process is still too imprecise to be effective. "Gene therapy is really a very blunt tool where we don't often control how much of the gene we get expressed. It's also variable from person to person. Gene doping would require significantly more precision than what the gene therapy tools would allow. So, it sounds like it should be super easy to do, but it's probably not."[63]

This hasn't stopped "biohackers" from trying. Among the most notorious of these "DIY biologists" is former NASA scientist Josiah Zayner, who publicly injected himself with DNA encoded for CRISPR that would, theoretically, modify his *MSTN* gene to enhance muscle growth.[64] It didn't work, but that didn't seem to matter. Zayner has since started a company that sells DIY CRISPR kits, including the one he used in his unsuccessful stunt. Such crude attempts recall early athletic experiments with brandy, egg whites, and strychnine (Chapter 5). Eventually, though, athletes will get it right and that, at least according to WADA, will put them in the wrong.

The question is how will anti-doping authorities detect genetic manipulation. Porteus is understandably reluctant to share the details about how this might happen, revealing only that gene doping may leave behind certain "signatures" in the body. These would appear "not just in your DNA, but the signatures could be in some other parts of the body that give away that you have undergone gene doping," he hedges. WADA is also working to develop "redundant detection methods so that if we miss with one, we pick it up with another."[65]

The Athlete Biological Passport offers another possibility for detecting changes in the body's chemistry (Chapter 4), but what if gene doping took place before an athlete reached the elite stage? It could even happen in childhood, if not in utero. In 2018, biophysicist He Jiankui announced that he had used CRISPR technology on embryos to produce twin girls who were resistant to HIV (the human immunodeficiency virus). His work was widely condemned, and Chinese authorities sentenced him to three years in prison for "illegal medical practice."[66] Still, the relative ease with which he manipulated embryonic DNA is cause for alarm. If parents are willing to genetically test their tender offspring, it is not a far leap to genetic manipulation. And if certain nations are already genetically testing citizens, and others engage in state-sponsored doping, the possibility of genetically modified Olympic teams seems entirely possible.

Genes Aren't Everything

By this point in the book, the arguments around gene-doping regulations should be fairly predictable. There is the unequivocal stance that gene doping is performance-enhancing, risks athletes' health, violates the spirit of sport, and should therefore remain banned. Even those athletes disinclined to dope will feel coerced if they want to be competitive.[67] There is also the grounded fear that doping at the elite level will trickle down to athletes at the lower levels.

Then there is the opposing libertarian view which holds that athletes should do with their bodies as they please. The only "sensible option" argues journalist Michael Le Page, is to "accept that there is no way to stop gene doping in the long run and reverse the 2003 ban on it before illicit and potentially dangerous forms of it become common."[68]

Finally, there are middle-ground positions, such as Miah's proposal to solve this "ethical dilemma" by forging "distinct, genetically enhanced competitions and to continue voluntary submission to anti-doping testing procedures."[69] This is the updated "parallel-federations" model—clean and gene doped—to which scholars Verner Møller and Rasmus Bysted Møller respond that the "only way this 'solves' the ethical dilemma is by throwing ethics out of the window." The doped category would transform sport into nothing more than "a gene-tech laboratory."[70]

Another centrist proposal is to allow gene doping under medical supervision and within certain parameters to maximize potential and attenuate genetic inequalities. This is in line with harm-reduction principles and what Savulescu calls "physiological doping" (Chapter 4).[71] Of course, this would also widen the striking and undeniable socioeconomic and technological divide in international sport. Then again, what doesn't?

Genes aren't everything, but they play an undeniable role in elite sport performance. We do not need studies to tell us that athletic talent is heritable, or something passed on genetically through family lineage.[72] Just look at American football's Manning family and their legacy of gifted quarterbacks, or tennis sisters Venus and Serena Williams, or, before them, Margaret and Romania Peters, who dominated the courts in the era of racial segregation. Olympians related to former Olympic medal winners have a greater probability of winning a medal of their own.[73] During the 2017 NBA finals, up to 40 percent of the players on the court at any time had a father who also played in the NBA.[74] As physiologist Per-Olof Astrand remarked in 1967, "anyone interested in winning Olympic gold medals must select his or her parents very carefully."[75]

All of this is, of course, much bigger than sport. At the same time, knowing about the genes of super-athletes might do a world of good. Researchers with Stanford University's ELITE project (Exercise at the Limit: Inherited Traits of Endurance) are studying the world's greatest endurance athletes, looking for candidates whose VO_2max falls within the top 0.02 percent of the population. "We are looking for the fittest people in the world," ELITE's recruiting website announces.[76] So far, their search has identified cyclists, rowers, and cross-country skiers as the fittest. Marathoners rarely make the cut.

The ELITE scientists are trying to find genetic variations that might account for an unusually high VO_2max, which could go a long way in understanding heart disease. "The better the heart, the more oxygen uptake," explains Dr. Mikael Mattsson, the project's managing investigator. "Find the genes that are important for high oxygen uptake, you find the genes that are important for a good heart."[77] If they can locate those genes, they can mimic their benefits with drugs or change them with gene therapy.

That pounding you hear is the crush of athletes and coaches clamoring at Mattsson's door.

Conclusion

The Promise of Sport

It took almost two years to resolve the 2022 Olympic team figure skating competition. The IOC delayed the medal ceremony—the first time in Olympic history for a completed event—while authorities debated what to do about Kamila Valieva's positive drug test (see Introduction). In early 2023, the Russian Anti-Doping Agency's (RUSADA) disciplinary tribunal determined that because Valieva was only fifteen years old at the time of the offense, she bore "no fault or negligence" for her use of trimetazidine. Her only penalty would be disqualification from the 2021 Russian National Championship, where she gave the tainted sample. There would be no additional sanction, which would award Valieva and her team the 2022 gold medal.[1]

In response, WADA and the ISU filed separate appeals to the Court of Arbitration for Sport. RUSADA's decision was "wrong under the terms of the World Anti-Doping Code," WADA argued.[2] According to the ISU, "all young athletes must be protected against doping. Such protection cannot happen by exempting young athletes from sanctions."[3] In other words, although Valieva was under the age of sixteen when she used performance-enhancing drugs, she should not be considered a "Protected Person," as defined in WADA's Code.

Finally, in January 2024, the Court of Arbitration for Sport sided with WADA and the ISU. A three-member arbitration panel disqualified

Valieva for four years, beginning with her 2021 positive doping test, and required her to forfeit "any titles, awards, medals, profits, prizes and appearance money" earned since then.[4] The ISU quickly voided Valieva's points from the 2022 Olympic team competition and demoted the Russians to a third-place finish.

In a media release, the Court of Arbitration for Sport explained that Valieva's rule violation was "one and the same whether the athlete is an adult or a Protected Person." There was "no basis" for treating her "any differently from an adult athlete."[5] She can return to competition just before the 2026 Winter Olympic Games. She will be nineteen years old. If seventeen is, as speculated, the "Tuberitze expiration date," one wonders what future Valieva has with her controversial coach or in sport altogether.

So who or what was ultimately protected in this case? Certainly not Valieva. To date, she has been the only person sanctioned for a transgression that was undoubtedly orchestrated by the adults in her life. Does the Court's decision protect other figure skaters, then? Figure skating itself? "All young athletes," as WADA contends? The elusive "spirit of sport?" There are no easy answers. Indeed, as explored throughout this book, protection is a slippery concept.

None of this is meant to suggest that protective policies have no place in sport. Rather, it is meant to ask how and why sport governing bodies create, implement, and enforce those policies, and the consequences—both intended and not—they have on the athletes they govern. Protective policies cannot fix elite sport's deep problems. Nothing short of a total demolition and rebuild can do that. However, there are steps elite sport governing bodies might take to build better protective policies.

Certainly, there are layers of institutional complexity, inside information, and bureaucratic entanglements that I cannot begin to unravel, but based on information from the previous chapters, as well as supplementary evidence, what follows are several possibilities that governing bodies might consider, moving forward. Specifically, policymakers could consider ways to make sport unexceptional, anticipate the possibilities for change, incorporate diverse and multidisciplinary insights, and center the athletes.

Make Sport Unexceptional

The protective policies explored in this book all rely, in one way or another, on sport exceptionalism, "the belief that sport is unique and requires its own special laws and rules," as sociologist Helen Lenskyj defines the term.[6] Elite sport is a transnational juggernaut with internal concerns that necessitate *lex sportiva* (sport law), but sport exceptionalism also allow an International Federation to stand as a "private body exercising private, contractual powers," rendering it exempt from national and international laws.[7] According to legal scholar Seema Patel, this positions "sports bodies outside of the legal regime and creates a gap in the protection of athletes' rights."[8] Put simply, sport exceptionalism benefits sport, not athletes.

One way to make sport unexceptional could be to align protective policies with human rights accords. This means, argue sociologists Bruce Kidd and Peter Donnelly, that governing bodies "ought to resort more systematically to the strategy of establishing, publicizing and drawing upon the charters, declarations and covenants that enshrine codes of entitlement and conduct."[9] This approach can extend throughout all aspects of sport's regulatory network. According to the Centre for Sport and Human Rights, the Court of Arbitration for Sport is "reluctant to factor human rights into its decisions" and its "arbitration rules do not offer adequate human rights protection."[10] What is more, writes legal scholar Lena Holzer, the court's arbitrators "have hardly any experience with human rights law."[11]

A human rights approach to sport governance cuts both ways. Human rights activists could recognize that sport is a meaningful aspect of people's lives—for better or worse. For this reason, sport could be made explicit in human rights documents, such as the United Nations' 1989 Convention of the Rights of the Child. One of nine core international human rights treaties, it is the most ratified human rights treaty in the world, and that which covers the widest range of issues. Yet there is no specific reference to sport. This is a significant oversight. Much of what likely happened to Valieva violates several articles in the convention, including those related to child labor, education, health, violence, abuse, exploitation, the best interests of the child, and "the illegal use of drugs."[12]

The perspective of human rights expert John Ruggie is useful in this regard. In 2015, the International Association Football Federation (Fédération Internationale de Football Association or FIFA) enlisted Ruggie to review its existing regulations and "publish a comprehensive and independent public report on what it means for FIFA to embed respect for human rights across the full range of its activities and relationships." In the end, Ruggie offered twenty-five recommendations for action, but he concluded with a caution:

> The foundational challenge for FIFA now is to go beyond putting words on paper and adding new administrative functions. What is required is a cultural shift that must affect everything FIFA does and how it does it. The result must be "good governance," not merely "good-looking governance."[13]

We can draw the same conclusions for protective policies. They must be "good policies," not merely "good-looking policies" that distract from or contribute to elite sport's deep problems.

Anticipating Change

Although implicitly understood, sport governing bodies could explicitly emphasize that protective policies are living documents that are subject to change as history shifts, knowledge evolves, and unintended consequences develop. WADA maintains that its Code "was never designed to be a document that stood still."[14] The IPC similarly "anticipates" that its classification code "will improve continuously, as will the ideas that would form rules, regulations, and policies that evolve with it."[15] Of course, there can be significant fallout when policies change, as when the World Para Athletics overhauled its classification system in 2018. Still, to openly address the possibilities for modification is to express a degree of humility and humanity that, frankly, elite sport too often lacks.

Planning for possible change involves an important distinction between unanticipated and unintended consequences. The former, argues political scientist Frank de Zwart, stems from "ignorance, error, or ideological blindness." The latter are the "unwelcome side effects that were foreseen but traded-off against intended consequences . . . and thus accepted."[16] Presumably, NCAA administrators recognized that sickle-cell trait is over-represented in Black athletes and that screening for it could therefore be discriminatory, but determined that safeguarding all athletes from the dangers of exertional sickling was more important (see Chapter 5).

It should go without saying that policymakers should (and hopefully do) anticipate all possible problems their decisions might precipitate. It also means that they should (but rarely do) acknowledge their complicity when problems inevitably arise. When, in 2022, the ISU opted to raise the age of senior eligibility to seventeen "for the sake of protecting the physical and mental health, and emotional well-being of Skaters," it failed to concede the ways that previous policy decisions encouraged the trend of young skaters who subsequently needed protection.[17]

Consider, as just one example, the ISU's 1988 decision to eliminate from competition the compulsory figures—those "painstaking tracings of variations of figure eights that required patience and maturity and gave the sport its name," describes journalist Christine Brennan.[18] Before 1988, junior skaters had to master the figures to advance to the senior ranks, which took time to accomplish and typically worked in favor of older athletes.

But as time wore on, the figures competition failed to make for "particularly exciting television broadcasts," assesses journalist Sandra Loosemore.[19] And so primarily for commercial reasons, the ISU dropped the figures from elite international competition. Insiders warned that the decision "would turn skating into jumping contests and might cause more injuries" and that "the sport would be dominated by fourteen and fifteen year old girls."[20] They were right: the surge in young skaters, difficult jumping, and injury rates in the sport were unintended but anticipated consequences. Arguably, then, the ISU prioritized its own interests over the health and well-being of its skaters, thus contributing to the need for its later age-based protective policy.

Diverse and Multidisciplinary Insights

Biocultural regulations must be grounded in sound, defensible, independent science. The goal ought to be to develop "evidenced-based policy," Lenskyj contends, "rather than relying on in-house experts to generate policy-based evidence."[21] This was not the case for World Athletics' 2018 Eligibility Regulations for the Female Classification. As detailed in Chapter 2, World Athletics funded the study and engaged federation-affiliated researchers to support its policy.[22] This is not just a conflict of interest. It is regulatory malfeasance.

Cultural insights are equally important to scientific evidence in establishing protective policies. There are historical, social, ethical, philosophical, legal, and organizational dimensions to legislating age, weight, sex, impairment, and enhancement. Creating policies that regulate bodies according to these biocultural categories therefore requires the collaboration of multidisciplinary teams of experts. No amount of scientific study, for instance, will stop athletes from engaging in dangerous weight-cutting tactics. These are cultural practices etched into sports like wrestling and ski jumping. And policies are meaningless when athletes compete as independent contractors without regulatory oversight, as they do in horse racing and mixed martial arts.

The multidisciplinary policymaking teams should also include experts from diverse social and geographic backgrounds. In crafting its 2021 Framework for Fairness, Inclusion and Non-Discrimination on the Basis of Gender Identity and Sex Variations, the IOC deliberately consulted a wide range of stakeholders, including a concerted effort to solicit the advice of representatives from the Global South (Chapter 2). This is important, not just because athletes from the region seem disproportionately affected by sex-testing policies, but also because the governance of elite sport has been—and continues to be—largely Western and Eurocentric and the legacies of colonialism and cultural imperialism continue to influence elite sport, its protective policies, and the athletes they govern.

Inclusion is also important for getting athletes and their local federations to buy into protective policies. According to scholars Jonathan Ruwuya, Byron Omwando Juma, and Jules Woolf, most African nations played "peripheral" roles in drafting WADA's anti-doping regulations. The "failure to adequately incorporate Africa," they argue, "was myopic and arguably led to decisions that undermined WADA's legitimacy in Africa." This has delayed the progress of anti-doping initiatives across the continent.[23]

Center the Athletes

Diverse and multidisciplinary policymaking teams should also include athletes. Athletes are at the center of protective policies, but as objects—not agents—of governance. As scientist Roger Pielke reasons, "the most compelling argument for athletes to have a greater voice in governing sport is that they are the stakeholders who are most directly affected by decisions about sport."[24] Philosopher Angela Schneider makes the same argument when it comes to constructing WADA's Prohibited List. "It's the athletes who take the risks and pay the price. They should decide what is on it."[25] Yet research shows that policymakers too often fail to collaborate with those who play the game.[26] Recall the ways that Para basketball included athletes in devising a new classification system, outlined in Chapter 3. As International Paralympic Committee President Philip Craven asserts, "Wheelchair basketball got its house in order by listening to the players and implementing what the majority of players thought was right."[27]

Centering the athlete in protective policies also involves implementing safe ways for them to report violations.[28] In this respect, child welfare is a particular cause for concern. According to one research team, "Sport has not yet developed a coherent global child protection, reporting, and response framework or system, which could potentially help promote, coordinate, and monitor sport safeguarding concerns in a similar way that the World Anti-Doping Agency addresses athlete doping."[29] WADA, to its credit, launched "Speak Up!" a digital platform for individuals to report—anonymously if so desired—allegations of anti-doping rule violations.[30] While there are national and sport-specific efforts to create similar monitoring systems for other types of endangerment and abuse, there is no unified, comprehensive, international system in place. It is incongruous to devote so much money and effort to anti-doping initiatives and not provide similar support to monitor, identify, and prevent athlete maltreatment.

To use WADA for a spark of inspiration, despite its many critiques, we might consider the agency's mandate for athletes' out-of-competition testing. As it professes, its "whereabouts rule" is "one of the most powerful means of deterrence and detection of doping."[31] Elite distance runner Kara Goucher recommends a comparable approach for monitoring athletes' health and well-being:

> There should be another independent body checking in on athletes, almost like antidoping. Not tied to any shoe brand or coach or governing body, just a safe place that checks in and makes sure that you are being treated OK. . . . We still need change when it comes to how we protect athletes.[32]

This is what athletic trainer Daniel Monthely has in mind when he advocates "spot checks" for dangerous weight-cutting practices in wrestling. Journalist Joan Ryan makes a similar point regarding Valieva's plight. She urges sport to open its "facilities to frequent and comprehensive inspection by child protective services provided by an international organization."[33]

If athletes agree to rules that subject them to unannounced, out-of-competition drug testing, similar rules should extend to the people and organizations charged with their care. The vast majority of elite sport takes place outside of organized contests—in places where athletes are sequestered and at their most vulnerable. The possibility of surprise inspections might discourage coaches and trainers from pushing deceptive practices like intentional misrepresentation or gene doping, or unhealthy methods for rapid weight loss.

In the end, these ideas, as well as others that flow from the previous chapters, will not correct sport's deep problems, but they may help us think about how we regulate bodies according to biocultural categories and just what it is that deserves protection. At the very least, I hope the contents of this book push us to think about what the policies mean, what they do, and who they most affect.

★★★★★

I have only had one recurring dream in my life, and it visits me every few months. In it, I learn the NCAA made a mistake—I have one more year of eligibility to play college soccer. I am thrilled. I am now fifty years old, would barely make it through warmups, and I am thrilled. And I am devastated every time I wake up and realize that the dream isn't real.

The truth is that I loved being an athlete. I loved it more than I loved anything else before I became a mother. I loved being a coach. I loved being a fan, a spectator, and a casual observer. Sport has given me an entire career.

Sport can be wonderful. Until it's not.

I previously wrote that I am not anti-sport and that I hoped to avoid the academic pitfalls of "sucking all the pleasure and fun" out of sport.[34] But a critical appraisal of protective policies has made me keenly aware of elite sport's deep problems. The Introduction to this book ended with a question: What are we willing to accept in the quest for *citius, altius, fortius*? If the rest of the book is any indication, the answer is: Quite a lot. Indeed, what I have learned while writing this book has made it difficult for me to love sport the same way that I used to. On darker days, I'm not so sure I love it at all.

I am not alone in this. For communications scholar David Heineman, to love sport was "to exercise purposeful, willful ignorance and indifference," to its deep problems.[35] "There are a lot of things that go on in sports that we may not agree with," admitted American football commentator Gina Wright. "If you literally had to not participate or support any sport because there is something you don't agree with, you probably wouldn't be a sports fan at all, let's be real."[36] So in the interest of being real, is it possible to reconcile fandom with sport's deep problems, or are we watching sport with our heads in the sand?

This is not an indictment of sporting pleasures, which are as true as they are transcendent. Rather, it is a condemnation of the ways that the elite sport industry erodes the promise of sport. It is a denouncement of the rapacious greed and the obscene amounts of money that could be used to develop sport in healthy, equitable, and meaningful ways, but instead line the pockets of government officials, sport executives, corporate sponsors, and cartel owners. It is a rebuke of the audiences who clamor for *citius, altius, fortius*, even as it jeopardizes athletes' health and well-being. None of this excuses the bad behavior of the athletes, but it does try to understand their deceptive practices within a system that exploits, dehumanizes, and commodifies them—that degrades their bodies, and minds, and spirits—that too often chews them up and spits them out.

It is also an indictment of sport's lofty banalities that, in the end, are little more than window dressing. There is a phenomenon known as sportswashing, in which an organization or political regime uses sport to sanitize its image. Qatar, for instance, has been accused of sportswashing its egregious record of human rights violations by hosting the

2022 FIFA men's World Cup, as Russia did before, as do any number of hosts and sponsors.

Sport is also guilty of sportswashing. It tries to cleanse its visage with sudsy ideals of Olympism, with rhetoric about "playing true" and the "spirit of sport," with hollow assertions that sport is a "human right" and healthful pursuit, and with tinny platitudes about level playing fields. At times, it feels as though protective policies are just another way of sportswashing too many deep problems.

But then, just when I am prepared to walk away from sport, a neighbor will give me tickets to a local women's basketball game. I will take my young daughters, and despite my mounting cynicism, I will find myself on the edge of my seat as our team ekes out a win, point by sweaty point, inch by hard-earned inch. I will yell, and groan, and pump my fists, and hold my breath, and curse the referee, and enjoy every single second of it.

We will sit behind a girls' league that will put on an exhibition at halftime, during which they will run their plays with studied concentration, and heave shots they have no business taking, and shake off their disappointment, and cheer each other on. When it is done, they will return to the stands with loose-jointed exuberance, and imagine themselves as the women on the court, and rewrite the stories they tell themselves about their halftime performance. They will leap to their feet for the television cameras, and dance for cameos on the jumbotron, and beg the cheerleaders to shoot at them with the t-shirt gun.

And then I will look to my left to see that my nine-year-old daughter is doing exactly the same thing. And I will look to my right to see that my typically disaffected eleven-year-old is out-leaping, out-dancing, out-begging them all. And I will remember: sport can be wonderful.

And it is worth protecting.

Notes

INTRODUCTION

1. Quoted in "Raising Competition Age for Figure Skaters Not Enough to Combat Abusive Coaches, Former Skaters Say," *CBC*, June 8, 2022, https://ca.news.yahoo.com/raising-competition-age-figure-skaters-003721081.html.

2. Bach quoted in Sean Ingle, "'Tremendous Coldness': IOC President Condemns Kamila Valieva's Entourage," *The Guardian,* February 18, 2022, https://www.theguardian.com/sport/2022/feb/18/tremendous-coldness-ioc-president-slams-kamila-valievas-entourage-over-skaters-treatment.

3. Kremlin spokesman Dmitry Peskov quoted in Daniel Chavkin, "Russia Responds to Criticism from IOC President About Valieva's Coach," *Sports Illustrated,* February 18, 2022, https://www.si.com/olympics/2022/02/18/kamila-valieva-coach-kremlins-spokesman-olympics-president. To her credit, Tutberidze coached Anna Shcherbakova and Alexandra Trusova to gold and silver in that same event.

4. Valieva's use of trimetazidine was discovered from an anti-doping test taken on December 25, 2021, at the Russian National Figure Skating Championships. The results were not determined until February 7, 2022, after Valieva had competed in the women's singles free skate component of the team event at the 2022 Olympic Games.

5. World Anti-Doping Agency, *An Athlete's Guide to the Significant Changes in the 2021 Code,* 2021, https://www.athleticsintegrity.org/downloads/pdfs/know-the-rules/en/Athlete-Guide-2021-Code_English_LIVE.pdf.

6. World Anti-Doping Agency, World Anti-Doping Code, January 1, 2021, https://www.wada-ama.org/en/resources/world-anti-doping-program/world-anti-doping-code. Initially, Valieva was provisionally suspended by the Russian Anti-Doping Agency (RUSADA) but she challenged the suspension, and it was lifted by RUSADA's disciplinary panel

following a hearing. On February 11 and 12, 2022, the International Olympic Committee (IOC), WADA, and the International Skating Union filed separate appeals to challenge the disciplinary panel's decision before the CAS Ad Hoc Division operating at the Olympic Games.

7. World Anti-Doping Agency, "WADA Statement on Court of Arbitration Decision to Declare Russian Anti-Doping Agency as Non-compliant," December 17, 2020, https://www.wada-ama.org/en/news/wada-statement-court-arbitration-decision-declare-russian-anti-doping-agency-non-compliant.

8. In 2018, the Dutch Skating Federation filed an "urgent proposal" with the ISU to raise the age of eligibility to seventeen; the Norwegian Skating Association submitted a similar proposal three years later. Michael Houston, "Norway Submit Figure Skating Age Limit Rise Despite Backlash," *Inside the Games,* November 30, 2020, https://www.insidethegames.biz/articles/1101434/norway-figure-skating-age-limit.

9. International Skating Union, *Agenda of the 58th Ordinary Congress, Phuket,* 2022, https://www.isu.org/docman-documents-links/isu-files/documents-communications/about-isu/congress-documents/28312-isu-communication-2472-1/file. The ISU adopted a gradual approach to increasing the age limit—to age sixteen for 2023–2024 and then to seventeen the following season. This applies to athletes in single and pair figure skating, synchronized skating, ice dance, and speed skating. See Sarah Teetzel, "Philosophic Perspectives on Doping Sanctions and Young Athletes," *Frontiers in Sports and Active Living* (2022), https://doi.org/10.3389/fspor.2022.841033.

10. Tara Lipinski (@taralipinski). "Raising the Age Limit," Instagram. Quoted in Ryan Glasspiegel, "Tara Lipinski Eviscerates New Figure Skating Age Limit: 'Broken System,'" *New York Post,* June 9, 2022, https://nypost.com/2022/06/09/tara-lipinski-eviscerates-new-figure-skating-age-limit/.

11. Lennard Davis and David Morris, "The Biocultures Manifesto," in *The End of Normal: Identity in A Biocultural Era,* ed. Lennard Davis (Ann Arbor: University of Michigan Press, 2014), 122.

12. See, as examples, Anne Fausto-Sterling, *Sex/gender: Biology in a Social World* (New York: Routledge, 2012); Katrina Karkazis, "The Misuses of 'Biological Sex,'" *The Lancet* 394, no. 10212 (2019): 1898–1899; L. Zachary DuBois and Heather Shattuck-Heidorn, "Challenging the Binary: Gender/Sex and the Bio-logics of Normalcy," *American Journal of Human Biology* 33, no. 5 (2021): e23623.

13. See Lindsay Parks Pieper, *Sex Testing: Gender Policing in Women's Sports* (Urbana: University of Illinois Press, 2016).

14. See Alexander B.T. McAuley, Joseph Baker, and Adam L. Kelly, "Defining 'Elite' Status in Sport: From Chaos to Clarity," *German Journal of Exercise and Sport Research* 52 (2022): 183–197.

15. Quoted in "One in Half-Million Chance of Making the Olympics," *Horsetalk*, July 25, 2012, https://www.horsetalk.co.nz/2012/07/25/one-in-half-million-chance-olympics/.

16. Allie Reynolds and Alireza Hamidian Jahromi, "Transgender Athletes in Sports Competitions: How Policy Measures Can Be More Inclusive and Fairer to All," *Frontiers in Sports and Active Living* 3 (2021), https://doi.org/10.3389/fspor.2021.704178.

17. I use "protectionism" as a gender-neutral alternative to "paternalism," or authoritative interventions that limit individual liberties for purportedly benevolent reasons. See Gerald Dworkin, "Paternalism," *The Monist* 56 (1972): 64–84. Anna Posberg also addresses protective policies in "Defining 'Woman': A Governmentality Analysis of How Protective Policies are Created in Elite Women's Sport," *International Review for the Sociology of Sport* (2022): 1–21. McAuley, Baker, and Kelly, "Defining 'Elite' Status in Sport."

18. World Para Swimming, "Classification in Para Swimming," accessed December 21, 2023, https://www.paralympic.org/swimming/classification.

19. "IAAF Publishes Briefing Notes and Q&A on Female Eligibility Regulations," May 7, 2019, https://worldathletics.org/news/press-release/questions-answers-iaaf-female-eligibility-reg.

20. WADA, World Anti-Doping Code. Emphasis in the original.

21. IOC, "Tokyo 2020 Event Programme To See Major Boost for Female Participation, Youth and Urban Appeal," June 9, 2017, https://www.olympic.org/news/tokyo-2020-event-programme-to-see-major-boost-for-female-participation-youth-and-urban-appeal.

22. International Olympic Committee, *Olympic Charter* (Lausanne: International Olympic Committee, 2021), 81.

23. Sarah Teetzel, "Minimum and Maximum Age Limits for Competing in the Olympic Games," in *Proceedings: International Symposium for Olympic Research* (London: Ontario: International Centre for Olympic Studies, 2010), 342.

24. Quoted in Philip Hersh, "Youth, Maturity—Gymnastics Needs Best of Both Worlds," *Chicago Tribune,* August 24, 1995, https://www.chicagotribune.com/news/ct-xpm-1995-08-24-9508240182-story.html.

25. Quoted in Alyssa Roenigk, "Karolyi Says Age Limit Would Rob Gymnasts of Golden Opportunity," *ESPN*, August 10, 2008, https://www.espn.com/olympics/summer08/gymnastics/columns/story?id=3527997.

26. United Nations, Convention on the Rights of the Child, 1989, https://www.ohchr.org/en/instruments-mechanisms/instruments/convention-rights-child.

27. Paulo David, *Human Rights in Youth Sport: A Critical Review of Children's Rights in Competitive Sport* (London: Routledge, 2004), 7. See also Siri Farstad, "Protecting Children's Rights in Sport: The Use of Minimum Age," *Human Rights Law Commentary* 3 (2007): 1–20.

28. Peter Donnelly, "Child Labour, Sport Labour: Applying Child Labour Laws to Sport," *International Review for the Sociology of Sport* 32, no. 4 (1997): 393.

29. Court of Arbitration for Sport, "Semenya, ASA and IAAF: Executive Summary," May 1, 2019, https://www.tas-cas.org/fileadmin/user_upload/CAS_Executive_Summary__5794_.pdf; Court of Arbitration for Sport, CAS 2020/A/6807 Blake Leeper v. International Association of Athletics Federations, accessed December 21, 2023, https://www.tas-cas.org/fileadmin/user_upload/Award__6807___for_publication_.pdf.

30. Helen Jefferson Lenskyj, *The Olympic Games: A Critical Approach* (Leeds: Emerald Insight, 2020), 108; Helen Jefferson Lenskyj, *Gender, Athletes' Rights, and the Court of Arbitration for Sport* (Leeds: Emerald Insight, 2018).

31. Quoted in William Fotheringham, *Put Me Back on My Bike: In Search of Tom Simpson* (New York: Random House, 2012), 169–170.

32. Mark A. Ware et al., "Cannabis and the Health and Performance of the Elite Athlete," *Clinical Journal of Sport Medicine* 28, no. 5 (2018): 480–484; Shgufta Docter et al., "Cannabis Use and Sport: A Systematic Review," *Sports health* 12, no. 2 (2020): 189–199.

33. International Olympic Committee, *Olympic Charter* (Lausanne: International Olympic Committee, 2021), 108. Emphasis added.

34. Lenskyj, *The Olympic Games*, 156.

35. Frank de Zwart, "Unintended but Not Unanticipated Consequences," *Theoretical Sociology* 44 (2015): 291.

36. Adam G. Pfleegor and Danny Roesenberg, "Deception in Sport: A New Taxonomy of Intra-lusory Guiles," *Journal of the Philosophy of Sport* 41, no. 2 (2014): 209–231.

37. Stanley W. Henson, "The Problem of Losing Weight," *Amateur Wrestling News*, February 12, 1969, 14; Ben Crighton, Graeme L. Close, and James P. Morton, "Alarming Weight Cutting Behaviours in Mixed Martial

Arts: A Cause for Concern and a Call for Action," *British Journal of Sports Medicine* 50, no. 8 (2016): 446–447.

38. "Crash Diets for Athletes Termed Dangerous," *Journal of the American Medical Association Health Bulletin*, February 1959, 9. Emphasis added.

39. Guilherme Giannini Artioli et al., "Authors' Reply to Davis: 'It Is Time to Ban Rapid Weight Loss from Combat Sports,'" *Sports Medicine* 47, no. 8 (2017): 1677–1681.

40. Ivan Waddington, "Theorising Unintended Consequences of Anti-Doping Policy," *Performance Enhancement and Health* 4, no. 3–4 (2016): 80–87.

41. Verner Møller, "The Road to Hell is Paved with Good Intentions—A Critical Evaluation of WADA's Anti-Doping Campaign," *Performance Enhancement and Health* 4, no. 3–4 (2016): 114.

42. "Raising the Age Limit for Skating Would End the Age of the Quad," *USA Today*, February 18, 2022, https://www.usatoday.com/story/spo rts/olympics/2022/02/18/raising-age-limit-for-skating-would-end-the-age-of-the-quad/49830469/.

43. Geoffroy Berthelot et al., "Has Athletic Performance Reached Its Peak?" *Sports Medicine* 45, no. 9 (2015): 1263–1271; Geoffroy Berthelot et al., "Athlete Atypicity on the Edge of Human Achievement: Performances Stagnate after the Last Peak in 1988," *PLoS ONE* 5 (2010):e8800 doi: 10.1371/journal.pone.0008800; Geoffroy Berthelot et al., "The Citius End: World Records Progression Announces the Completion of a Brief Ultra-Physiological Quest," *PLoS ONE* 3 (2008):e1552 doi: 10.1371/ journal.pone.0001552; Mark W. Denny, "Limits to Running Speed in Dogs, Horses and Humans," *Journal of Experimental Biology* 211 (2008), 3836–3849; Thomas Hagen, Espen Tønnessen, and Stephen Seiler, "9.58 and 10.49: Nearing the Citius End for 100 m?" *International Journal of Sports Physiology and Performance* 10 (2015): 269–272; Alan M. Nevill and Gregory Whyte, "Are There Limits to Running World Records?" *Medicine and Science in Sports and Exercise* 37, no. 10 (2005): 1785–1788; Alan M. Nevill et al., "Are There Limits to Swimming World Records?" *International Journal of Sports Medicine* 28, no. 12 (2007): 1012–1017.

44. See, for example, Giuseppe Lippi et al., "Updates on Improvement of Human Athletic Performance: Focus on World Records in Athletics," *British Medical Bulletin* 87 (2008): 7–15; Nigel Balmer, Pascoe Pleasence, and Alan Nevill, "Evolution and Revolution: Gauging the Impact of Technological and Technical Innovation on Olympic Performance," *Journal of Sports Science* 30 (2011): 1075–1083.

45. Court of Arbitration for Sport, *Arbitral Award Delivered by the Court of Arbitration for Sport, CAS 2018/O/5794, Mokgadi Caster Semenya v. International Association of Athletics Federations; CAS 2018/O/5798 Athletics South Africa v. International Association of Athletics Federations*, 2019, 160.

46. Denhollander quoted in Juliet Macur, "A Spectacle That Shook the World of Skating," *New York Times,* February 19, 2022.

47. See Christopher C. Grenfell and Robert E. Rinehart, "Skating on Thin Ice: Human Rights in Youth Figure Skating," *International Review for the Sociology of Sport* 38, no. 1 (2003): 79–97; Jennifer Lipetz and Roger J. Kruse, "Injuries and Special Concerns of Female Figure Skaters," *Clinics in Sports Medicine* 19, no. 2 (2000): 369–380; Moa Jederström et al., "468 Determinants of Sports Injury in Young Female Swedish Competitive Figure Skaters," *British Journal of Sports Medicine* 55, Suppl 1 (2021): A179–A179; Agnieszka D. Kowalczyk et al., "Pediatric and Adolescent Figure Skating Injuries: A 15-year Retrospective Review," *Clinical Journal of Sport Medicine* 31, no. 3 (2019): 295–303.

48. Michael A. Messner, *Taking the Field: Women, Men, and Sports* (Minneapolis: University of Minnesota Press, 2002), 76.

49. Franklin Foer, "The Goals of Globalization," *Foreign Policy* 12, no. 5 (2005), https://foreignpolicy.com/2009/10/20/the-goals-of-globalization/.

50. Alan Bairner, "An Introduction to the Negative Aspects of Sport," *Idrotts Forum,* August 29, 2022, https://idrottsforum.org/baiala_anderson-magrath220829/.

51. Quoted in Verity Bowman and Luke Mintz, "The Dark Truth behind the Beauty of Figure Skating," *Daily Telegraph,* February 11, 2022, Gale in Context database.

52. Quoted in Anna Moeslein, "Watch Kamila Valieva Become the First Woman to Land a Quad at the Olympics," *Glamour,* February 7, 2022, https://www.glamour.com/story/watch-kamila-valieva-become-the-first-woman-to-land-a-quad-at-the-olympics.

CHAPTER 1

1. Lee appeared on Chael Sonnen's *You're Welcome* podcast, quoted in Mike Chiari, "Kevin Lee Says He Didn't Know Where He Was during UFC 216 Weight Cut," *Bleacher Report,* October 12, 2017, https://bleacherreport.com/articles/2738367-kevin-lee-says-he-didnt-know-where-he-was-during-ufc-216-weight-cut.

2. Quoted in Dave Doyle, 'Devastated' Kevin Lee Says Weight Cut 'Damn Near Killed Me,'" October 8, 2017, mmafighting.com/2017/10/8/16443192/devastated-kevin-lee-says-weight-cut-damn-near-killed-me.

3. The same is true for athletes who put on prodigious weight, as in sumo wrestling and American football.

4. Morteza Khodaee et al., "Rapid Weight Loss in Sports with Weight Classes," *Current Sports Medicine Reports* 14, no. 6 (2015): 435–441.

5. John Connor and Brendan Egan, "Comparison of Hot Water Immersion at Self-Adjusted Maximum Tolerance Temperature, with or without the Addition of Salt, for Rapid Weight Loss in Mixed Martial Arts," *Biology of Sport* 38, no. 1 (2021): 89–96.

6. Simon Rottenberg, "The Baseball Players' Labor Market," *Journal of Political Economy* 64 (1956): 242–258.

7. Louis M. Burke et al., "ACSM Expert Consensus Statement of Weight Loss in Weight-Category Sports," *Current Sports Medicine Reports* 20, no. 4 (2021): 199–271.

8. Analiza M. Silva, Diana A. Santos, and Catarina N. Matias, "Weight-Sensitive Sports," in *Body Composition: Health and Performance in Exercise and Sport*, ed. Henry C. Lukaski (Boca Raton, FL: CRC Press, 2017): 233–284; Jorunn Sundgot-Borgen and Ina Garthe, "Elite Athletes in Aesthetic and Olympic Weight-Class Sports and the Challenge of Body Weight and Body Compositions," *Journal of Sport Sciences* 29, Suppl 1 (2011): S101–S114.

9. Nicola Ialongo, Raphael Hermann, and Lorenz Rahmstorf, "Bronze Age Weight Systems as a Measure of Market Integration in Western Eurasia," *PNAS* 118, no. 27 (2021): e2105873118; C. St. C. B. Davison, "Landmarks in the History of Weighing and Measuring," *Transactions of the Newcomen Society* 31, no. 1 (1957): 131–152.

10. Quoted in Ian Whitelaw, *A Measure of All Things: The Story of Man and Measurement* (New York: St. Martin's Press, 2007), 20.

11. In 1960, the metric system was renamed as the International System of Units (SI), which has seven base units: meter, kilogram, second, ampere, kelvin, candela, and mole.

12. Quoted in Henri Moreau, "The Genesis of the Metric System and the Work of the International Bureau of Weights and Measures," *Journal of Chemical Education* 30, no. 1 (1953): 5.

13. Peter N. Stearns, *Fat History: Bodies and Beauty in the Modern West* (New York: NYU Press, 2002), 27.

14. Hillel Schwartz, *Never Satisfied: A Cultural History of Diets, Fantasies and Fat* (New York: The Free Press, 1986), 153–159; Amanda M. Czerniawski,

"From Average to Ideal: The Evolution of the Height and Weight Table in the United States, 1836–1943," *Social Science History* 31, no. 2 (2007): 273–296.

15. Donald G. Kyle, "Greek Athletic Competitions: The Ancient Olympics and More," in *A Companion to Sport and Spectacle in Greek and Roman Antiquity*, ed. Paul Christesen and Donald G. Kyle (Chichester: Wiley Blackwell, 2013), 29.

16. Allen Guttmann, *From Ritual to Record: The Nature of Modern Sports* (New York: Columbia University Press, 1978), 60.

17. Tony Collins, *Sport in a Capitalist Society: A Short History* (Routledge, 2013), 2.

18. Neil Tranter, *Sport, Economy and Society in Britain 1750–1914* (Cambridge: Cambridge University Press, 1998).

19. Roger Longrigg, *The History of Horse Racing* (New York: Stein and Day, 1972), 39.

20. Wray Vamplew, *The Turf: A Social and Economic History of Horse Racing* (London: Allen Lane, 1976), 17.

21. James Rice, *History of the British Turf, from the Earliest Times to the Present Day*, vol. II (London: Sampson Low, Marston, Searle, and Rivington, 1879), 367. "Handicap" is derived from the phrase "hand in the cap" or "hand i' cap," a practice used for drawing lots. See G. Herbert Stutfield, "Handicaps," *National Review* 31, no. 183 (May 1898): 391; Jørn Hansen, "The Origins of the Term Handicap in Games and Sports—History of a Concept," *Physical Culture and Sport Studies Research* 65 (2015): 7–13.

22. See "light-weight | light weight, n. and adj." *OED Online*, accessed June 8, 2021, Oxford University Press, https://www.oed.com/view/Entry/108247?rskey=1EHf1H&result=1&isAdvanced=false; "heavy-weight, n." OED Online, accessed June 8, 2021, Oxford University Press. https://www.oed.com/view/Entry/85259?redirectedFrom=heavyweight. Related terms entered the lexicon in the early 1800s, including "feather weight," or the "lightest weight allowed by the rules to be carried by a horse in a handicap. Hence sometimes applied to the rider," as the *Oxford English Dictionary* reports.

23. Henry John Rous, *On the Laws and Practice of Horse Racing* (London: A. H. Baily, 1866), 12–13; Wray Vamplew, *The Turf*, 119.

24. Wray Vamplew, "Playing with the Rules: Influences on the Development of Regulation in Sport," *International Journal of the History of Sport* 24, no. 7 (2007): 864.

25. Irene McCanliss, *Weight on the Thoroughbred Horse* (Chester, MA, privately printed, 1967), 112.

26. See Walter Gilbey, *Horses Past and Present* (London:Vinton & Co., 1900), available at https://www.gutenberg.org/files/43580/43580-h/43580-h.htm (p. 51).

27. Mike Huggins, "Racing Culture, Betting, and Sporting Protomodernity: The 1975 Newmarket Carriage Match," *Journal of Sport History* 42, no. 2 (2015): 326.

28. C. M. Prior, *Early Records of the Thoroughbred Horse* (1924), quoted in McCanliss, *Weight on the Thoroughbred Horse*, 25.

29. Collins, *Sport in a Capitalist Society*, 8.

30. Wray Vamplew, "Reduced Horse Power: The Jockey Club and the Regulation of British Horseracing," *Entertainment Law* 2, no. 3 (2003): 94–95.

31. McCanliss, *Weight on the Thoroughbred Horse*, 30.

32. Vamplew, "Reduced Horse Power," 95. The Jockey Club first made weighing in compulsory in 1875.

33. Historically, young children, decidedly smaller than average adults. In connection to Chapter 1, the regular use of child jockeys contributed to the introduction of age limits in horse racing and camel racing, although there are still places in the world where the practice continues. See Nicholas McGeehan, "Spinning Slavery: The Role of the United States and UNICEF in the Denial of Justice for the Child Camel Jockeys of the United Arab Emirates," *Journal of Human Rights Practice* 5, no. 1 (2013): 96–124.

34. Kasia Boddy, "'Under the Queensberry Rules, So to Speak': Some Versions of a Metaphor," *Sport in History*, 31, no. 4 (2011): 400. Fighting cocks were also weighed for the purpose of competition during this time. Iris M. Middleton, "Cockfighting in Yorkshire during the Early Eighteenth Century," *Northern History* 40, no. 1 (2003): 129–146.

35. Elliott J. Gorn, "The Bare-Knuckle Era," in *The Cambridge Companion to Boxing*, ed. Gerald Early (Cambridge: Cambridge University Press, 2019), 38.

36. Elliott J. Gorn, *The Manly Art: Bare-Knuckle Prize Fighting in America* (Ithaca, NY: Cornell University Press, 1986).

37. Jeffrey Sammons, *Beyond the Ring: The Role of Boxing in American Society* (Urbana: University of Illinois Press, 1988), 13–14; 19; S. Derby Gisclair, *The Olympic Club of New Orleans: Epicenter of Professional Boxing, 1883–1897* (Jefferson, NC: McFarland, 2018), 18.

38. W. Russel Gray, "For Whom the Bell Tolled, The Decline of British Prize Fight in the Victorian Era," *Journal of Popular Culture* 21, no. 2 (1987): 59.

39. Kenneth G. Sheard, "Aspects of Boxing in the Western Civilizing Process," *International Review for the Sociology of Sport* 32, no. 1 (1997): 36.

40. "The 'Police Gazette' Champion Belt, *National Police Gazette*, November 1, 1884, 13.

41. Gorn, *The Manly Art*, 242.

42. Benjamin G. Rader, *American Sports: From the Age of Folk Games to the Age of Televised Sports*, 3rd ed. (Englewood Cliffs, NJ: Prentice Hall, 1996), 46.

43. Jack Anderson, "A Brief Legal History of Prize Fighting in Nineteenth Century America," *Sport in History* 24, no. 1 (2004): 61.

44. Robert G. Rodriguez, *The Regulation of Boxing* (Jefferson, NC: McFarland, 2009), 30.

45. Jennifer McClearen, *Fighting Visibility: Sports Media and Female Athletes in the UFC* (Urbana: University of Illinois Press, 2021), 35.

46. Raúl Sánchez García and Dominic Malcolm, "Decivilizing, Civilizing or Informalizing? The International Development of Mixed Martial Arts," *International Review for the Sociology of Sport* 45, no. 1 (2010): 39–58.

47. Robert J. Szczerba, "Mixed Martial Arts and the Evolution of John McCain," *Forbes,* April 3, 2014, https://www.forbes.com/sites/roberts zczerba/2014/04/03/mixed-martial-arts-and-the-evolution-of-john-mccain/?sh=7ff53dc42d59.

48. Ben Crighton, Graeme L. Close, and James P. Morton, "Alarming Weight Cutting Behaviours in Mixed Martial Arts: A Cause for Concern and a Call for Action," *British Journal of Sports Medicine* 50, no. 8 (2016): 446–447.

49. Oliver R. Barley, Dale W. Chapman, and Chris R. Abbiss, "Weight Loss Strategies in Combat Sports and Concerning Habits in Mixed Martial Arts," *International Journal of Sports Physiology and Performance* (2017): 1–24.

50. Crighton, Close, and Morton, "Alarming Weight Cutting," 446–447.

51. Quoted in McClearen, *Fighting Visibility*, 99–100.

52. Jason Duncan, "A Re-Union for MMA: Reoccurring Issues Plaguing Mixed Marshal Arts Fighters and Potential Solutions," *University of Denver Sports and Entertainment Law Journal* 23 (2020): 21.

53. Henson, "The Problem of Losing Weight," 14.

54. Morteza Khodaee et al., "Rapid Weight Loss in Sports with Weight Classes," *Current Sports Medicine Reports* 14, no. 6 (2015): 435–441; Ciro José Brito et al., "Methods of Body Mass Reduction by Combat Sport Athletes," *International Journal of Sport Nutrition and Exercise Metabolism* 22 (2012): 89–97; Guilherme Giannini Artioli et al., "Prevalence, Magnitude,

and Methods of Rapid Weight Loss Among Judo Competitors," *Medicine and Science in Sports and Exercise* 42 (2010): 436–442.

55. Joseph John Matthews and Ceri Nicholas, "Extreme Rapid Weight Loss and Rapid Weight Gain Observed in UK Mixed Martial Arts Athletes Preparing for Competition," *International Journal of Sport Nutrition and Exercise Metabolism* 27, no. 2 (2017): 122–129.

56. Adam M. Jetton et al., "Dehydration and Acute Weight Gain in Mixed Martial Arts Fighters Before Competition," *Journal of Strength and Conditioning Research* 27, no. 5 (2013): 1322–1326.

57. Mike Bohn, "UFC's Cris 'Cyborg': Why She's Forced to Lose Weight to Win," *Rolling Stone,* September 22, 2016, https://www.rollingstone. com/culture/culture-sports/ufcs-cris-cyborg-why-shes-forced-to-lose-weight-to-win-187532/.

58. Paul T. von Hippel, Caroline G. Rutherford, and Katherine M. Keyes, "Gender and Weight Among Thoroughbred Jockeys: Underrepresented Women and Underweight Men," *Socius* 3 (2017): 1–7; Philippa Velija and Leah Flynn, "'Their Bottoms are the Wrong Shape': Female Jockeys and the Theory of Established Outsider Relations," *Sociology of Sport Journal* 27, no. 3 (2010): 301–315; "Trainers 'Do No Use Female Jockeys' Despite Findings of New Study," *BBC,* January 30, 2018, https://www. bbc.com/sport/horse-racing/42870971; https://www.telegraph.co.uk/ news/2018/01/30/women-jockeys-just-good-men-finds-14-year-review/.

59. Joel M. Press et al., "The National Jockey Injury Study: An Analysis of Injuries to Professional Horse-racing Jockeys," *Clinical Journal of Sport Medicine* 5, no. 4 (1995): 236.

60. Scott A. Gruender, *Jockey: The Rider's Life in American Thoroughbred Racing* (Jefferson, NC: McFarland and Company, 2007), 78–79.

61. Demete Labadarios, Juan Kotze, D. Momberg, and T. V. W. Kotze, "Jockeys and Their Practices in South Africa," *Nutrition and Fitness for Athletes* 71 (1993): 97–114; Mark A. Leydon and Clare Wall, "New Zealand Jockeys' Dietary Habits and Their Potential Impact on Health," *International Journal of Sport Nutrition and Exercise Metabolism* 12, no. 2 (2002): 220–237; Jan M. Moore et al., "Weight Management and Weight Loss Strategies of Professional Jockeys," *International Journal of Sport Nutrition and Exercise Metabolism* 12, no. 1 (2002): 1–13; Martin Tolich and Martha Bell, "The Commodification of Jockeys' Working Bodies: Anorexia or Work Discipline," in *A Global Racecourse: Work, Culture, and Horse Sports*, ed. Chris McConville (Melville: Australian Society for Sport History, 2008), 101–133; Eimear Dolan et al., "Nutritional,

Lifestyle, and Weight Control Practices of Professional Jockeys," *Journal of Sports Sciences* 29, no. 8 (2011): 791–799; Press et al., "The National Jockey Injury Study," 236–240.

62. Quoted in Bill Vourvoulias, "In Kentucky Derby, Race Against Weight Takes Jockeys to a Dark Side," *Fox News,* May 2, 2014, http://www.foxnews.com/sports/2014/05/02/in-kentucky-derby-race-against-weight-takes-jockeys-to-dark-side.html.

63. George Wilson et al., "GB Apprentice Jockeys Do Not Have the Body Composition to Make Current Minimum Race Weights: Is It Time to Change the Weights or Change the Jockeys?" *International Journal of Sport Nutrition and Exercise Metabolism* 1, no. 2 (2020): 101–104.

64. Valerie DeBenedette, "For Jockeys, Injuries Are Not a Long Shot," *The Physician and Sportsmedicine* 15, no. 6 (1987): 236–245; Press et al., "The National Jockey Injury Study," 236–240; Kylie Legg et al., "Jockey Career Length and Risk Factors for Loss from Thoroughbred Race Riding," *Sustainability* 12, no. 18 (2020): 7443; Wray Vamplew, "Still Crazy After All Those Years: Continuity in a Changing Labour Market for Professional Jockeys," *Sport in Society* 19, no. 3 (2016): 378–399.

65. Gruender, *Jockey*, 78.

66. Ibid., 91.

67. Mike Chapman, *Encyclopedia of American Wrestling* (Champaign, IL: Human Kinetics, 1990), 78.

68. William J. Monilaw, "The Effects of Training Down in Weight on the Growing Boy and How to Control or Abolish the Practice," *The School Review* 25, no. 5 (1917): 350–360.

69. W. W. Tuttle, "The Effect of Weight Loss by Dehydration and the Withholding of Food on the Physiologic Responses of Wrestlers," *Research Quarterly* 14 (1943): 158–166. See also H. E. Kenney, "The Problem of Making Weight for Wrestling Meets," *Journal of Health and Physical Education* 1 (1930): 49; Olden Curtice Gillum, "The Effects of Weight Reduction on the Bodily Strength of Wrestlers," PhD diss., Ohio State University, 194); Nathan Doscher, "The Effect of Rapid Weight Loss upon the Performance of Wrestlers and Boxers and upon the Physical Proficiency of College Students," *Research Quarterly in Exercise and Sport* 15 (1944): 317–324; Vernon Ekfelt, "Eliminating the Criticisms of High School Wrestling," *Athletic Journal* (December 1955), 10–11.

70. Philip J. Rasch and Walter Kroll, *What the Research Tells the Coach about Wrestling* (Washington, DC: American Association for Health, Physical Education, and Recreation, 1964), 41.

71. Tim Noakes, *Waterlogged: The Serious Problem of Overhydration in Endurance Sports* (Champaign, IL: Human Kinetics, 2012), xiii.

72. Belgian cyclist-turned-trainer George Ronsse, quoted in William Fotheringham, *Put Me Back on My Bike: In Search of Tom Simpson* (London: Yellow Jersey Press, 2007), 180.

73. Jaime Schultz, W. Lawrence Kenney, and Andrew D. Linden, "Heat-Related Deaths in American Football: An Interdisciplinary Approach," *Sport History Review* 45, no. 2 (2014): 123–144; Steven R. Murray and Brian E. Udermann, "Fluid Replacement: A Historical Perspective and Critical Review," *International Sports Journal* (2003): 59–73.

74. Kathleen Bachynski, *No Game for Boys to Play: The History of Youth Football and the Origins of a Public Health Crisis* (Chapel Hill: University of North Carolina Press, 2019); Jack W. Berryman, *Out of Many, One: A History of the American College of Sports Medicine* (Champaign, IL: Human Kinetics, 1995); William Bock, "The Effects of Dehydration upon the Cardio-Respiratory Endurance of Wrestlers," PhD diss., Ohio State University, 1965, 1–3; Louis Elfenbaum, "The Physiological Effects of Rapid Weight Loss Among Wrestlers," PhD diss., Ohio State University, 1966; William D. Paul, "Crash Diets and Wrestling," *Journal of the Iowa Medical Society* 56, no. 8 (1966): 835–840; Charles M. Tipton and Tse-Kia Tcheng, "Iowa Wrestling Study: Weight Loss in High School Students," *JAMA* 214, no. 7 (1970): 1269–1274.

75. American Medical Association, "Wrestling and Weight Control."

76. "Doctors Suggest Wrestling Ban," *Philadelphia Inquirer*, December 17, 1967, 13. See Charles M. Tipton, Tse-Kia Tcheng, and W. D. Paul, "Evaluation of the Hall Method for Determining Minimum Wrestling Weights," *Journal of the Iowa Medical Society* 59 (1969): 571–574.

77. American College of Sports Medicine "Position Stand on Weight Loss in Wrestlers," *Medicine and Science in Sports and Exercise* 8, no. 2 (1976): xi–xiii; Paul M. Ribisi and William G. Herbert, "Effects of Rapid Weight Reduction and Subsequent Rehydration upon the Physical Working Capacity of Wrestlers," *Research Quarterly* 41, no. 4 (1970): 536–41; Norman C. Hansen, "Wrestling with "Making Weight," *Physician and Sportsmedicine* 6 (1978): 106–111.

78. Interview with Daniel Monthley, September 29, 2021.

79. Rob Prebish, *The Solitary Wrestler: Methods for Safe Weight Control* (JBE Online Books, 2006), http://www.jbeonlinebooks.org/wrestling/documents/SolitaryWrestler-Chapter04.pdf.

80. Suzanne Nelson Steen and Kelly D. Brownell, "Patterns of Weight Loss and Regain in Wrestlers: Has the Tradition Changed?" *Medicine and Science in Sports and Exercise* 22, no. 6 (1990): 762–768.

81. The wrestlers were nineteen-year-old Billy Saylor, a freshman at Campbell University; twenty-two-year-old Joseph LaRosa, a senior at the University of Wisconsin-La Crosse; and twenty-one-year-old Jeff Reese, a junior at the University of Michigan.

82. The NCAA based a number of these changes on the Wisconsin's plan for high school wrestlers. See Robert A. Opplinger et al., "The Wisconsin Wrestling Minimal Weight Project: A Model for Wrestling Weight Control," *Medicine and Science in Sports and Exercise* 27, no. 8 (1995): 1220–1224; Rebecca L. Carl, Miriam D. Johnson, and Thomas J. Martin, "Promotion of Healthy Weight-Control Practices in Young Athletes," *Pediatrics* 140, no. 3 (2017): e20171871; doi: https://doi.org/10.1542/peds.2017-1871.

83. Jack Ransone and Brian Hughes, "Body-Weight Fluctuations in Collegiate Wrestlers: Implications of the National Collegiate Athletic Association Weight-Certification Program," *Journal of Athletic Training* 39, no. 2 (2004): 162–165.

84. Robert A. Opplinger, Suzanne A. Nelson Steen, and James R. Scott, "Weight Loss Practices of College Wrestlers," *International Journal of Sport Nutrition and Exercise Metabolism* 13, no. 1 (2003): 29–46; Aimee E. Gibbs, Joel Pickerman, and Jon K. Sekiya, "Weight Management in Amateur Wrestling," *Sports Health* 1, no. 3 (2009): 227–230.

85. Liz Neporent, "Winter Olympic Sports: One Size Doesn't Fit All," *ABC News,* February 11, 2014, https://abcnews.go.com/Sports/winter-olympic-sports-size-fits/story?id=22447486.

86. Bernhard Schmölzer and Wolfram Müller, "The Importance of Being Light: Aerodynamic Forces and Weight in Ski Jumping," *Journal of Biomechanics* 35, no. 8 (2002): 1059–1069.

87. "Ski Jumpers Are Taking the Light Approach," *Washington Times*, February 9, 2002, https://www.washingtontimes.com/news/2002/feb/9/20020209-035615-1739r/.

88. Quoted in Philip Hersh, "Are Ski Jumpers Too Thin?" *Chicago Tribune,* January 16, 2002, http://articles.chicagotribune.com/2002-01-16/sports/0201160397_1_top-jumpers-alan-alborn-sven-hannawald.

89. Wolfram Müller, Dieter Platzer, and Bernhard Schmölzer, "Dynamics of Human Flight on Skis: Improvement on Safety and Fairness in Ski Jumping," *Journal of Biomechanics* 29, no. 8 (1996): 1061–1068.

90. International Ski Federation, Rule 1.2.1.1, "Ski Length," in *Specifications for Competition Equipment, Edition 2018/2019* (Oberhofen, Switzerland: International Ski Federation, 2018), 6–7.

91. Wolfram Müller et al., "Underweight in Ski Jumping: The Solution to the Problem," *International Journal of Sports Medicine* 27, no. 11 (2005): 926–934; Jeré Longman, "Battle of Weight Versus Gain in Ski Jumping," *New York Times,* February 11, 2010, https://www.nytimes.com/2010/02/12/sports/olympics/12skijump.html?mtrref=www.google.com&gwh=6C5ADE01DF8E74AEA0C123D0BFC8CB3E&gwt=pay.

92. Mikko Virmavirta and Juha Kivekäs, "Is It Still Important to Be Light in Ski Jumping?" *Sports Biomechanics* 20, no. 4 (2021): 407–418.

93. Farzin Halabchi, "Doping in Combat Sports," in *Combat Sports Medicine,* ed. Ramin Kordi et al. (London: Springer, 2009), 55–72.

94. Petterson, Ekström, and Berg, "Practices of Weight Regulation"; Steen and Brownell, "Patterns of Weight Loss and Regain."

95. Natalie Voss, "'They Just Want to Ride': Small Changes in Scale of Weights Have Big Impact on Jockeys' Health," *Paulick Report,* February 8, 2017, https://www.paulickreport.com/news/ray-s-paddock/just-want-ride-small-changes-scale-weights-big-impact-jockeys-health/; Dubnov-Raz Gal et al., "Can Height Categories Replace Weight Categories in Striking Martial Arts Competitions? A Pilot Study," *Journal of Human Kinetics* 47 (2015): 91; Ana De la Fuente García, "Height Categories as a Healthier Alternative to Weight Categories in Taekwondo Competition," *Revista de Artes Marciales Asiáticas* 13, no. 1 (2018): 53–60; Luca Paolo Ardigò, Wissem Dhahbi, and Johnny Padulo, "Height-Based Model for the Categorization of Athletes in Combat Sports," *European Journal of Sport Science* (2020): 1–10.

96. James MacDonald, "MMA: The Dangers of Cutting Weight in Mixed Martial Arts," *Bleacher Report,* accessed August 28, 2021, https://bleacherreport.com/articles/1487089-mma-the-dangers-of-cutting-weight-in-mixed-martial-arts.

CHAPTER 2

1. Quinn, who helped the Canadian women's soccer team win its first gold medal, also identifies as transgender and nonbinary. BMX rider Chelsea Wolfe was an alternate for the US team. In this chapter, I use the terms "trans" and "transgender" interchangeably. The Human Rights Campaign defines "transgender" as "an umbrella term for people whose gender identity and/or expression is different from cultural

expectations based on the sex they were assigned at birth." Additionally, "Intersex people are born with a variety of differences in their sex traits and reproductive anatomy. There is a wide variety of difference among intersex variations, including differences in genitalia, chromosomes, gonads, internal sex organs, hormone production, hormone response, and/or secondary sex traits." "Differences of sex development" is a medical term often associated with intersexuality. Human Rights Campaign, "Glossary of Terms," accessed December 21, 2021, https://www.hrc.org/resources/glossary-of-terms.

2. International Olympic Committee, IOC Consensus Meeting on Sex Reassignment and Hyperandrogenism, November 2015, https://stillmed.olympic.org/Documents/Commissions_PDFfiles/Medical_commission/2015-11_ioc_consensus_meeting_on_sex_reassignment_and_hyperandrogenism-en.pdf.

3. Lance Wahlert and Autumn Fiester, "Gender Transports: Privileging the 'Natural' in Gender Testing Debates for Intersex and Transgender Athletes," *American Journal of Bioethics* 12, no. 7 (2012): 20.

4. Court of Arbitration for Sport, "Semenya, ASA and IAAF: Executive Summary," May 1, 2019, https://www.tas-cas.org/fileadmin/user_upload/CAS_Executive_Summary__5794_.pdf.

5. Ibid.

6. International Olympic Committee, *Olympic Charter* (Lausanne: IOC, 2021), 8.

7. Human Rights Watch, "End Abusive Sex Testing for Women Athletes," December 4, 2020, https://www.hrw.org/news/2020/12/04/end-abusive-sex-testing-women-athletes#.

8. "IAAF Publishes Briefing Notes and Q&A on Female Eligibility Regulations," May 7, 2019, https://www.worldathletics.org/news/press-release/questions-answers-iaaf-female-eligibility-reg.

9. Ali Durham Greey and Helen Jefferson Lenskyj, "Conclusion: Challenges, Struggles and the Way Forward," in *Justice for Trans Athletes*, ed. Ali Durham Greey and Helen Jefferson Lenskyj (Leeds: Emerald Insight, 2023), 165.

10. Valérie Thibault et al., "Women and Men in Sport Performance: The Gender Gap Has Not Evolved Since 1983," *Journal of Sports Science and Medicine* 9, no. 2 (2010): 214–223; Jonathan Ospina Betancurt et al., "Hyperandrogenic Athletes: Performance Differences in Elite-Standard 200m and 800m Finals," *Journal of Sports Sciences* 36, no. 21 (2018): 2464–2471.

11. Eric Dunning, "Sport as a Male Preserve: Notes on the Social Sources of Masculine Identity and Its Transformations," *Theory, Culture & Society* 3, no. 1 (1986): 79–90.

12. Michael A. Messner, "Sports and Male Domination: The Female Athlete as Contested Ideological Terrain," *Sociology of Sport Journal* 5, no. 3 (1988): 197–211; Gail Bederman, *Manliness and Civilization: A Cultural History of Gender and Race in the United States, 1880–1917* (Chicago: University of Chicago Press, 2008).

13. Jaime Schultz, *Qualifying Times: Points of Change in U.S. Women's Sport* (Urbana: University of Illinois Press, 2014); Susan K. Cahn, *Coming on Strong: Gender and Sexuality in Women's Sport* (Urbana: University of Illinois Press, 2015).

14. Jaime Schultz, *Women's Sports: What Everyone Needs to Know* (New York: Oxford University Press, 2018), 92.

15. Frederick Rogers Rand, "Olympics for Girls?" *School & Society* 30 (August 1929): 194.

16. Lindsay Parks Pieper, "The Medical Examination of Lady Competitors: Sex Control in Skiing, 1967–2000," *International Journal of the History of Sport* 10, no. X (2021): 1–19.

17. Grantland Rice, "Separate Olympics for Sexes in 1940 Planned," *Los Angeles Times,* August 12, 1936, A9.

18. Quoted in Dennis J. Frost, *Seeing Stars: Sports Celebrity, Identity, and Body Culture in Modern Japan* (Cambridge, MA: Harvard University Press, 2010), 143.

19. "Olympic Games," *Time,* August 10, 1936, 28.

20. Clare Tebbutt, "The Spectre of the 'Man-woman Athlete': Mark Weston, Zdenek Koubek, the 1936 Olympics and the Uncertainty of Sex," *Women's History Review* 24, no. 5 (2015): 721–738; Sonja Erikainen, "The Story of Mark Weston: Re-centering Histories and Conceptualising Gender Variance in 1930s International Sport," *Gender and History* 32, no. 2 (2020): 304–319.

21. "Helen Stephens A Man, Polish Writer Thinks," *Chicago Tribune,* August 6, 1936, 20. See also Sharon Kinney Hanson, *The Life of Helen Stephens: The Fulton Flash* (Carbondale: Southern Illinois University Press, 2004), 96.

22. Sonja Erikainen, "Hybrids, Hermaphrodites, and Sex Metamorphoses: Gendered Anxieties and Sex Testing in Elite Sport, 1937–1968," in *Gender Panic, Gender Policy*, ed. Vasilikie Demos and Marcia Texler Segal (Leeds: Emerald Insight, 2017), 167.

23. IAAF, *Handbook of the International Amateur Athletic Federation* (1937), quoted in Erikainen, "The Story of Mark Weston," 315.

24. Vanessa Heggie, "Testing Sex and Gender in Sports; Reinventing, Reimagining and Reconstructing Histories," *Endeavour* 34, no. 4 (2010): 157–163.

25. "Sex Test Disqualifies Athlete," *New York Times*, September 16, 1967, 28.

26. Heggie, "Testing Sex."

27. "Chromosomes Do Not an Athlete Make," *JAMA*, 202, no. 11 (1967): 54.

28. Monique Berlioux, "Femininity," *Olympic Review* 3 (December 1967): 1–2.

29. Ibid.

30. Mary Peters with Ian Wooldridge, *Mary P.: Autobiography* (London: Stanley Paul, 1974), 55–56.

31. "Preserving la Difference," *Time*, September 16, 1966, 74.

32. Quoted in Lindsay Parks Pieper, *Sex Testing: Gender Policing in Women's Sports* (Urbana: University of Illinois Press, 2016), 54.

33. Fiona Alice Miller, "'Your True and Proper Gender': The Barr Body as a Good Enough Science of Sex," *Studies in History and Philosophy of Science Part C: Studies in History and Philosophy of Biological and Biomedical Sciences* 37, no. 3 (2006): 459–483.

34. Daniel F. Hanley quoted in "Chromosomes Do Not an Athlete Make," *JAMA,* 202, no. 11 (1967): 55.

35. Eduardo Hay, "Sex Determination in Putative Female Athletes," *JAMA* 221, no. 9 (1972): 999; Vanessa Heggie, "Subjective Sex: Science, Medicine and Sex Tests in Sports," in *Transgender Athletes in Competitive Sport*, ed. Eric Anderson and Ann Travers (London: Taylor & Francis, 2017), 131–142.

36. "Sex Test Disqualifies Athlete," *New York Times* September 16, 1967, 28; Joe Leigh Simpson et al., "Gender Verification in Competitive Sports," *Sports Medicine* 16, no. 5 (1993): 305–315.

37. "Records of Polish Girl Sprinter Who Flunked Sex Test Barred," *New York Times*, February 26, 1968, 50.

38. Dr. Clayton Thomas, quoted in Deborah Larned, "The Femininity Test: A Woman's First Olympic Hurdle," *Womensports*, July 1976, 10.

39. Hay, "Sex Determination," 998.

40. Keith Moore, "Sexual Identity of Athletes, *JAMA* 205 (1968): 163–164.

41. Quoted in Pieper, *Sex Testing*, 67.

42. Los Angeles Olympic Organization Committee, Games of the XXIIIrd Olympiad, Los Angeles 1984, International Olympic medical controls brochure, 1980, http://www.la84foundation.org.

43. Jonathan Ospina-Betancurt, Eric Vilain, and María José Martinez-Patiño, "The End of Compulsory Gender Verification: Is It Progress for Inclusion of Women in Sports?" *Archives of Sexual Behavior* (September 2021): 1–9.

44. María José Martínez-Patiño, "Personal Account: A Woman Tried and Tested," *Lancet* 366 (December 2005): S38.

45. Ibid.

46. Ibid.

47. Quoted in Alison Carlson, "When Is a Woman Not a Woman?" *Women's Sport & Fitness*, March 1991, 29.

48. Martínez-Patiño, "Personal Account," S38.

49. For this reason, I have opted not to include photographs of women affected by sex-testing policies.

50. In 1991, IAAF council replaced "gender verification" with "a medical examination for the health and well-being of all athletes (women and men) that would "obviate the need for any laboratory-based genetic 'sex test.'" The IAAF dropped the medical examination to following year, determining that observations during urinalyses for doping tests were sufficient for detecting male interlopers in women's sports. Arne Ljungqvist and Joe Leigh Simpson, "Medical Examination for Health of All Athletes Replacing the Need for Gender Verification in International Sports: The International Amateur Athletic Federation Plan," *JAMA* 267, no. 6 (1992): 850–852.

51. Bernard Dingeon, "Gender Verification and the Next Olympic Games," *JAMA* 269, no. 3 (1993): 357–358.

52. Louis J. Elsas, Risa P. Hayes, and Kasinathan Muralidharan, "Gender Verification at the Centennial Olympic Games," *Journal of the Medical Association of Georgia* 86, no. 1 (1997): 50.

53. Quoted in Gail Vines, "Last Olympics for the Sex Test?" 41.

54. Peters and Wooldridge, *Mary P.*, 57.

55. Janet Heinonen, "A Decent Proposal: Keeping Track," *International Track and Field Newsletter*, March 1994, 24.

56. Louis J. Elsas et al., "Gender Verification of Female Athletes," *Genetics in Medicine* 2, no. 4 (2000): 249–254. See also Berit Skirstad, "Gender Verification in Competitive Sport: Turning from Research to Action," in *Values in Sport: Elitism, Nationalism, Gender Equality and the Scientific Manufacture of Winners*, ed. Torbjörn Tännsjö and Claudio Marcello Tamburrini (London: Taylor & Francis, 2000): 116–122; Andy Brown, "DSD & Transgender Athletes: Paula Radcliffe's View," *Sports Integrity Initiative*, April 30, 2019, https://www.sportsintegrityinitiative.com/

dsd-transgender-athletes-paula-radcliffes-view/;   Doriane   Lambelet
Coleman, "Sex, Sport, and Why Track and Field's New Rules on
Intersex Athletes Are Essential, *New York Times*, April 30, 2018, https://
www.nytimes.com/2018/04/30/sports/track-gender-rules.html.

57. IAAF Policy on Gender Verification (2006).

58. Brenna Munro, "Caster Semenya: Gods and Monsters," *Safundi* 11, no. 4
(2010): 383–396; Neville Hoad, "Run, Caster Semenya, Run! Nativism
and the Translations of Gender Variance," *Safundi* 11, no. 4 (2010): 397–
405; Zine Magubane, "Spectacles and Scholarship: Caster Semenya,
Intersex Studies, and the Problem of Race in Feminist Theory," *Signs* 39,
no. 3 (2014): 761–785; Tavia Nyong'o, "The Unforgivable Transgression
of Being Caster Semenya," *Women & Performance: A Journal of Feminist
Theory* 20, no. 1 (2010): 95–100; Kathryn Henne and Madeleine Pape,
"Dilemmas of Gender and Global Sports Governance: An Invitation
to Southern Theory," *Sociology of Sport Journal* 35, no. 3 (2018): 216–
225; Katrina Karkazis and Rebecca M. Jordan-Young, "The Powers of
Testosterone: Obscuring Race and Regional Bias in the Regulation
of Women Athletes," *Feminist Formations* 30, no. 2 (2018): 1–39; Krystal
Batelaan and Gamal Abdel-Shehid, "On the Eurocentric Nature of Sex
Testing: The Case of Caster Semenya," *Social Identities* 27, no. 2 (2021):
146–165.

59. "Indian Runner Fails Gender Test, Loses Medal," ESPN.com. December
18, 2006, http://sports.espn.go.com.

60. Quoted in Nilanjana Bhowmick and Jyoti Thottam, "Gender and
Athletics: India's Own Caster Semenya," *Time*, September 1, 2009,
http://content.time.com/time/world/article/0,8599,1919562,00.html;
Hameet Shah Singh, "India Athlete Makes Plea for Semenya," *CNN,*
September 14, 2009, http://www.cnn.com/2009/WORLD/asiapcf/
09/14/Semenya.India.Athlete/index.html. See also Payoshni Mitra,
"The Untold Stories of Female Athletes with Intersex Variations in
India," in *Routledge Handbook of Sport, Gender and Sexuality*, ed. Eric
Anderson and Jennifer Hargreaves (London: Routledge, 2014), 384–394.

61. Schultz, *Qualifying Times*, 118.

62. "Semenya 'Maybe Not 100pc' a Woman," *ABC News*, September 11,
2009, http://www.abc.net.au/news/2009-09-11/semenya-maybe-not-
100pc-a-woman/1424994.

63. Jaime Schultz, "Caster Semenya and the 'Question of too': Sex Testing
in Elite Women's Sport and the Issue of Advantage," *Quest* 63, no. 2
(2011): 228–243.

64. IAAF, "Caster Semenya May Compete," July 6, 2010, http://iaaf.org.

65. IAAF Regulations Governing Eligibility of Females with Hyperandrogenism to Compete in Women's Competition (2011).

66. International Olympic Committee, IOC Regulations on Female Hyperandrogenism (2012).

67. International Association of Athletics Federations, IAAF Regulations Governing Eligibility of Females with Hyperandrogenism to Compete in Women's Competition (2011), https://www.worldathletics.org/news/iaaf-news/iaaf-to-introduce-eligibility-rules-for-femal-1.

68. Court of Arbitration for Sport, *Interim Arbitral Award: Dutee Chand v. Athletics Federation of India and the International Association of Athletics Federations, CAS 2014/A/3759* (July 24, 2015); Stephen Lazarou, Louis Reyes-Vallejo, and Abraham Morgentaler, "Wide Variability in Laboratory Reference Values for Serum Testosterone," *Journal of Sex Medicine* 3, no. 6 (2006): 1085–1089; Angelica Lindén Hirschberg, "Hyperandrogenism in Female Athletes," *Journal of Clinical Endocrinology and Metabolism* 104, no. 2 (2019): 503–505; Richard V. Clark et al., "Large Divergence in Testosterone Concentrations between Men and Women: Frame of Reference for Elite Athletes in Sex-Specific Competition in Sports, a Narrative Review," *Clinical Endocrinology* 90, no. 1 (2019): 15–22.

69. Katrina Karkazis et al., "Out of Bounds? A Critique of the New Policies on Hyperandrogenism in Elite Female Athletes," *American Journal of Bioethics* 12, no. 7 (2012): 3–16. Emphasis in original.

70. Stéphane Bermon et al., "Are the New Policies on Hyperandrogenism in Elite Female Athletes Really Out of Bounds?" *American Journal of Bioethics* 13, no. 5 (2013): 63–65. For conflicting opinions regarding testosterone and performance, see Marie-Louise Healy et al., "Endocrine Profiles in 693 Elite Athletes in the Postcompetition Setting," *Clinical Endocrinology* 81, no. 2 (2014): 294–305; Stéphane Bermon et al., "Serum Androgen Levels in Elite Female Athletes," *Journal of Clinical Endocrinology and Metabolism,* 99, no. 11 (2014): 4328–4335; David J. Handelsman, Angelica L. Hirschberg, and Stéphane Bermon, "Circulating Testosterone as the Hormonal Basis of Differences in Athletic Performance," *Endocrine Reviews* 39, no. 5 (2018): 803–882; Rebecca Jordan-Young and Katrina Karkazis, *Testosterone: An Unauthorized Biography* (Cambridge, MA: Harvard University Press, 2019); Cara Tannenbaum and Sheree Bekker, "Sex, Gender, and Sports," *British Medical Journal* 364 (2019): l1120;

71. Jaime Schultz, "Good Enough? The 'Wicked' Use of Testosterone for Defining Femaleness in Women's Sport," *Sport in Society* 24, no. 4 (2021): 607–627.

72. IAAF Regulations Governing Eligibility of Females with Hyperandrogenism to Compete in Women's Competition (2011).

73. Karkazis and Jordan-Young, "The Powers of Testosterone."

74. Patrick Fénichel et al., "Molecular Diagnosis of 5α-reductase Deficiency in 4 Elite Young Female Athletes through Hormonal Screening for Hyperandrogenism," *Journal of Clinical Endocrinology & Metabolism* 98, no. 6 (2013): E1055–E1059. Emphasis added.

75. Peter Sönksen et al., "Medical and Ethical Concerns Regarding Women with Hyperandrogenism and Elite Sport," *Journal of Clinical Endocrinology and Metabolism* 100, no. 3 (2015): 825–827.

76. Annet Negesa, "I Cannot Go Back to the Body I Had Before I Was Operated on, but I Can Try to Stop Other Women Going Through What I Did," *Telegraph,* November 18, 2021, https://www.telegraph.co.uk/athletics/2021/11/18/annet-negesa-cannot-go-back-body-had-operated-can-try-stop-women/#:~:text=Cheltenham%20Festival-,Annet%20Negesa%3A%20'I%20cannot%20go%20back%20to%20the%20body%20I,going%20through%20what%20I%20did'&text = Nine%20years%20ago%20sports%20authorities,feminine%2C%E2%80%9D%20or%20quit%20running. See also Human Rights Watch, *They're Chasing Us Away from Sport: Human Rights Violations in Sex Testing of Elite Athletes,* December 4, 2020, https://www.hrw.org/report/2020/12/04/theyre-chasing-us-away-sport/human-rights-violations-sex-testing-elite-women#.

77. *CAS 2014/A/3759,* 9–10.

78. *CAS 2014/A/3759* 59; 36. See also Richard J. Auchus, "Endocrinology and Women's Sports: The Diagnosis Matters," *Law and Contemporary Problems* 80, no. 127 (2018): 135–136.

79. *CAS 2014/A/3759,* 154. Emphasis in original.

80. Ross Tucker, "Hyperandrogenism and Women vs Women vs Men in Sport: A Q&A with Joanna Harper," *Science of Sport,* May 23, 2016, https://sportsscientists.com/2016/05/hyperandrogenism-women-vs-women-vs-men-sport-qa-joanna-harper/.

81. The IAAF's 2018 Eligibility Regulations for the Female Classification (Athletes with Differences of Sex Development) listed "relevant" DSDs as 5α-reductase type 2 deficiency; partial androgen insensitivity syndrome; 17β-hydroxysteroid dehydrogenase type 3 (17β-HSD3) deficiency; congenital adrenal hyperplasia; 3β-hydroxysteroid dehydrogenase deficiency; ovotesticular DSD; or any other genetic

disorder involving disordered gonadal steroidogenesis. In 2019, the regulations were updated to omit congenital adrenal hyperplasia and 3β-hydroxysteroid dehydrogenase deficiency.

82. Stéphane Bermon et al., "Serum Androgen Levels Are Positively Correlated with Athletic Performance and Competition Results in Elite Female Athletes," *BJSM* 52, no. 23 (2018): 1531–1532.

83. Peter H. Sönksen et al., "Hyperandrogenism Controversy in Elite Women's Sport: An Examination and Critique of Recent Evidence," *British Journal of Sports Medicine* 52, no. 23 (2018): 1481–1482; Simon Franklin, Jonathan Ospina Betancurt, and Silvia Camporesi, "What Statistical Data of Observational Performance Can Tell Us and What They Cannot: The Case of Dutee Chand v. AFI & IAAF," *British Journal of Sports Medicine* 52, no. 7 (2018): 420–421; Amanda Menier, "Use of Event-Specific Tertiles to Analyse the Relationship between Serum Androgens and Athletic Performance," *British Journal of Sports Medicine* 52, no. 23 (2018): 1540; Roger Pielke, Ross Tucker, and Erik Boye, "Scientific Integrity and the IAAF Testosterone Regulations," *International Sports Law Journal* 19, no. 1 (2019): 18–26; Roger Pielke, "Caster Semenya Ruling: Sports Federation Is Flouting Ethics Rules," *Nature,* May 17, 2019, https://www.nature.com/articles/d41586-019-01606-8.

84. "Correction: Serum Androgen Levels and Their Relation to Performance in Track and Field: Mass Spectrometry Results from 2127 Observations in Male and Female Elite Athletes," *British Journal of Sports Medicine* 55 (2021): e7.

85. Court of Arbitration for Sport, *Arbitral Award Delivered by the Court of Arbitration for Sport, CAS 2018/O/5794, Mokgadi Caster Semenya v. International Association of Athletics Federations; CAS 2018/O/5798 Athletics South Africa v. International Association of Athletics Federations* (2019), 17–18, https://www.tas-cas.org/fileadmin/user_upload/CAS_Award_-_redacted_-_Semenya_ASA_IAAF.pdf.

86. Ibid., 9–10.

87. Ibid., 133; 71–72.

88. Ibid., 68.

89. Julian Savulescu, "Ten Ethical Flaws in the Caster Semenya Decision on Intersex in Sport," May 9, 2019, https://theconversation.com/ten-ethical-flaws-in-the-caster-semenya-decision-on-intersex-in-sport-116448.

90. CAS, *Arbitral Award*, 160.

91. Interview with Payoshni Mitra, December 8, 2021.

92. After puberty, the healthy adult male produces 15 to 20 percent more circulating testosterone than the healthy adult, premenopausal woman. This partly contributes to the average man, relative to the average woman, developing larger and stronger bones, greater muscle mass and strength, and higher circulating hemoglobin, all of which can contribute to sport performance. Handelsman et al., "Circulating Testosterone as the Hormonal Basis of Differences in Athletic Performance."

93. David Handlesman, email to Eric Vilain, cited in CAS, *Arbitral Award*, 36.

94. CAS, *Arbitral Award*, 91.

95. World Athletics, "World Athletics Council Decides on Russia, Belarus and Female Eligibility," March 23, 2023, https://worldathletics.org/news/press-releases/council-meeting-march-2023-russia-belarus-female-eligibility.

96. Lindsay Parks Pieper, "Gender Regulation: Renée Richards Revisited," *International Journal of the History of Sport* 29, no. 5 (2012): 675–690.

97. Arne Ljungqvist and Myron Genel, "Transsexual Athletes—When Is Competition Fair?" *Lancet* 366 (2005): S42–43.

98. Elsas et al., "Gender Verification of Female Athletes," 250.

99. Ljungqvist and Genel, "Transsexual Athletes."

100. Statement of the Stockholm Consensus on Sex Reassignment in Sports (2003), https://stillmed.olympic.org/Documents/Reports/EN/en_report_905.pdf; Bethany Alice Jones et al., "Sport and Transgender People: A Systematic Review of the Literature Relating to Sport Participation and Competitive Sport Policies," *Sports Medicine* 47, no. 4 (2017): 701–716.

101. Sheila L. Cavanaugh and Heather Sykes, "Transsexual Bodies at the Olympics: The International Olympic Committee's Policy on Transsexual Athletes at the 2004 Athens Summer Games," *Body & Society* 12, no. 3 (2006): 89.

102. World Athletics' 2019 Eligibility Regulations for Transgender Athletes stipulated that trans women must suppress their endogenous testosterone below 5 nmol/L, in accordance with its DSD policy. IAAF, "Decisions Made at IAAF Council Meetings in Doha," October 14, 2019, https://www.worldathletics.org/news/press-release/iaaf-council-219-decisions.

103. World Anti-Doping Agency, TUE Physician Guidelines, Medical Information to Support the Decisions of TUE Committees: Female-to-Male (FtM) Transsexual Athletes (March 2016), https://www.wada-ama.org. Emphasis added.

104. World Anti-Doping Agency, TUEC Guidelines: Medical Information to Support the Decisions of TUE Committees—Transgender Athletes (October 20, 2017), https://www.wada-ama.org/en/resources/science-medicine/medical-information-to-support-the-decisions-of-tuecs-female-to-male. See Kristen Worley and Johanna Schneller, *Woman Enough: How a Boy Became a Woman and Changed the World of Sport* (Toronto: Vintage, 2020).

105. World Anti-Doping Agency, TUEC Guidelines: Medical Information to Support the Decisions of TUE Committees—Transgender Athletes (October 20, 2017), https://www.wada-ama.org/en/resources/science-medicine/medical-information-to-support-the-decisions-of-tuecs-female-to-male.

106. International Olympic Committee, IOC Framework on Fairness, Inclusion and Non-Discrimination on the Basis of Gender Identity and Sex Variations, 2021, https://stillmed.olympics.com/media/Docume nts/News/2021/11/IOC-Framework-Fairness-Inclusion-Non-dis crimination-2021.pdf?_ga=2.116948229.2094909257.1637082260-499116176.1634933505.

107. Ryan Storr, Madeleine Pape, and Sheree Bekker, "A Win for Transgender Athletes and Athletes with Sex Variations," *The Conversation,* November 18, 2021, https://theconversation.com/a-win-for-transgender-athle tes-and-athletes-with-sex-variations-the-olympics-shifts-away-from-testosterone-tests-and-toward-human-rights-172045?utm_source=facebook&utm_medium=bylinefacebookbutton&fbclid= IwAR36WYmbOdtU7UBwxj3b1w9ZEkgJyuXKClKFXLiBn14co 2rDYT5k-8SmG-Q.

108. Quoted in Katie Barnes, "IOC Provides Framework for International Federations to Develop Their Own Eligibility Criteria for Transgender, Intersex Athletes," *ESPN.com,* November 16, 2021, https://www.espn. com/olympics/story/_/id/32645620/ioc-provides-framework-intern ational-federations-develop-their-own-eligibility-criteria-transgender-intersex-athletes.

109. Joanna Harper et al., "Implications of a Third Gender for Elite Sports," *Current Sports Medicine Reports* 17, no. 2 (2018): 42–44.

110. Eileen McDonagh and Laura Pappano, *Playing with the Boys: Why Separate Is Not Equal in Sports* (New York: Oxford University Press, 2009).

111. Andria Bianchi, "Something's Got to Give: Reconsidering the Justification for a Gender Divide in Sport," *Philosophies* 4, no. 2 (2019): https://doi.org/10.3390/philosophies4020023.

112. Vikki Krane and Heather Barber, "Creating a New Sport Culture: Reflections on Queering Sport," in *Sex, Gender, and Sexuality in Sport: Queer Inquiries*, ed.Vikki Krane (London: Routledge, 2019), 231. See also Mary Jo Kane, "Resistance/Transformation of the Oppositional Binary: Exposing Sport as a Continuum," *Journal of Sport and Social Issues* 19, no. 2 (1995): 191–218; Dee Amy-Chinn, "The Taxonomy and Ontology of Sexual Difference: Implications for Sport," *Sport in Society* 15, no. 9 (2012): 1291–1305; Catherine Jean Archibald, "Transgender and Intersex Sports Rights," *Virginia Journal of Social Policy & Law* 26, no. 3 (2019): 246–276.

113. Bennett Foddy and Julian Savulescu, "Time to Re-evaluate Gender Segregation in Athletics?" *British Journal of Sports Medicine* 45, no, 15 (2011): 1184–1188.

114. Taryn Knox, Lynley C. Anderson, and Alison Heather, "Transwomen in Elite Sport: Scientific and Ethical Considerations," *Journal of Medical Ethics* 45, no. 6 (2019): 401.

115. Maayan Sudai, "The Testosterone Rule—Constructing Fairness in Professional Sport," *Journal of Law and Biosciences* 4, no. 1 (2017): 181–193; Irena Martínková, "Unisex Sports: Challenging the Binary," *Journal of the Philosophy of Sport* 47, no. 2 (2020): 248–265; Roslyn Kerr and C. Obel, "Reassembling Sex: Reconsidering Sex Segregation Policies in Sport," *International Journal of Sport Policy and Politics* 10, no. 2 (2018): 305–320.

CHAPTER 3

1. Sarah Dhanaphatana, "Freshman Swimmer Beats the Odds," *Daily Trojan* (University of Southern California), September 28, 2014, https://dailytrojan.com/2014/09/28/freshman-swimmer-beats-the-odds/.

2. World Para Swimming, "Classification in Para Swimming," accessed December 22, 2023, https://www.paralympic.org/swimming/classification.

3. Ibid.

4. IPC Swimming, Rules and Regulations, 2011, 57, https://www.paralympic.org/sites/default/files/document/120706163426076_2011_05_30__Swimming_Classification_Regulations.pdf.

5. Ibid.

6. Para swimming has additional classifications for athletes who compete in the breaststroke and the individual medley.

7. Quoted in Amy Rosewater, "Are Officials Redefining the Paralympian?" *ESPN,* July 15, 2015, http://www.espn.com/olympics/story/_/id/13258441/paralympics-officials-redefining-paralympian.

8. Ian Silverman, "Ian Silverman Writes Letter to IPC about Classification Issues," *SwimSwam,* July 31, 2016, https://swimswam.com/paralympic-champ-ian-silverman-writes-letter-to-ipc-about-fraud/.

9. In alignment with the World Health Organization, the IPC advises using the term "impairment" instead of disability. Officials argue that the word "disability" carries negative connotations—with the "dis" prefix implying a lack of something—in this case, a lack of ability, which mischaracterizes both high-performance athletes and a range of other individuals. "Impairment," the IPC contends, "shifts the focus more onto athletes' abilities and that they are able to achieve." International Paralympic Committee, "Guide to Reporting on Persons with an Impairment," October 2014, https://m.paralympic.org/sites/default/files/document/141027103527844_2014_10_31+Guide+to+reporting+on+persons+with+an+impairment.pdf.

10. Silverman, "Ian Silverman."

11. World Para Swimming, "World Para Swimming to Introduce Revised Classification Rules and Regulations from 2018," September 29, 2017, https://www.paralympic.org/news/world-para-swimming-introduce-revised-classification-rules-and-regulations-2018.

12. "20-Year Old Para-Swimmer Ailbhe Kelly Announces Retirement," *SwimSwam,* September 15, 2019, https://swimswam.com/20-year-old-para-swimmer-ailbhe-kelly-announces-retirement/.

13. Amy Marren, Facebook Post, February 1, 2020, https://www.facebook.com/amy.marren/posts/10222643039731150.

14. See Brendan Burkett et al., "Performance Characteristics of Para Swimmers: How Effective Is the Swimming Classification System?," *Physical Medicine and Rehabilitation Clinics* 29, no. 2 (2018): 333–346; Luca Puce et al., "Impact of the 2018 World Para Swimming Classification Revision on the Race Results in International Paralympic Swimming Events," *German Journal of Exercise and Sport Research* 50, no. 2 (2019): 1–13.

15. Quoted in Veronica Allan, "Do Para-Athletes Face Abuse Trying to Prove Their Disabilities," *News Decoder,* March 13, 2019, https://news-decoder.com/2019/03/13/para-athletes-classification-abuse/.

16. International Paralympic Committee, IPC Athlete Classification Code (Bonn: International Paralympic Committee, 2015), https://www.paralympic.org/sites/default/files/2020-05/170704160235698_2015_12_17%2BClassification%2BCode_FINAL2_0-1.pdf.

17. Sean M. Tweedy, Mark J. Connick, and Emma M. Beckman, "Applying Scientific Principles to Enhance Paralympic Classification Now and in the Future: A Research Primer for Rehabilitation Specialists," *Physical Medicine and Rehabilitation Clinics of North America* 29 (2018): 314.

18. P. David Howe, "From Inside the Newsroom: Paralympic Media and the 'Production' of Elite Disability," *International Review for the Sociology of Sport* 43, no. 2 (2008): 148.

19. John R. Gold and Margaret M. Gold, "Access for All: The Rise of the Paralympic Games," *Perspectives in Public Health* 127, no. 3 (2007): 122–141; Jason Bantjes and Leslie Swartz, "Social Inclusion through Para Sport: A Critical Reflection on the Current State," *Physical Medicine and Rehabilitation Clinics of North America* 29 (2018): 409–416; Chui Ling Goh, "To What Extent Does the Paralympic Games Promote the Integration of Disabled Persons into Society?" *International Sports Law Journal* 20 (2020): 36–54.

20. World Health Organization, *World Report on Disability* (Geneva: WHO Press, 2011), 3.

21. United Nations, "Factsheet on Persons with Disabilities," accessed June 25, 2021, https://www.un.org/development/desa/disabilities/resources/factsheet-on-persons-with-disabilities.html.

22. Simon Darcy, "The Paralympic Movement: A Small Number of Behemoths Overwhelming a Large Number of Also-Rans—A Pyramid Built on Sand," in *The Palgrave Handbook of Paralympic Studies*, ed. Ian Brittain and Aaron Beacom (London: Palgrave Macmillan, 2018), 234.

23. Andrew Novak, "Disability Sport in Sub-Saharan Africa: From Economic Underdevelopment to Uneven Empowerment," *Disability and the Global South* 1, no. 1 (2014): 44–63. See also Caroline Buts et al., "Socioeconomic Determinants of Success at the Summer Paralympics," *Journal of Sports Economics* 14, no. 2 (2011): 133–147.

24. Leslie Swartz et al., "'A More Equitable Society': The Politics of Global Fairness in Paralympic Sport," *PLOS One* 11, no. 12 (2016): e0167481, doi: 10.1371/journal.pone.0167481.

25. Gwen Knapp, "Why Ukraine's Small Paralympic Team Packs Such a Big Punch," *New York Times,* September 6, 2021. Russian athletes technically competed for the Russian Paralympic Committee, due to the four-year ban from WADA (see Introduction).

26. World Bank Country and Lending Groups, accessed September 16, 2021, https://datahelpdesk.worldbank.org/knowledgebase/articles/906519-world-bank-country-and-lending-groups.

27. Steve Bailey, *Athlete First: A History of the Paralympic Movement* (Hoboken, NJ: Wiley, 2008); Ian Brittain, *The Paralympic Games Explained* (London: Routledge, 2016); Karen P. Depauw and Susan J. Gavron, *Disability Sport*, 2nd ed. (Champaign, IL: Human Kinetics, 2005); Karen P. DePauw, "A Historical Perspective of the Paralympic Games," *Journal of Physical Education, Recreation & Dance* 83, no. 3 (2012): 21–31.

28. Throughout this chapter, I use "people-first language" as opposed to "identity-first language," with the understanding that there are important, if contested, cultural and political differences between the two. See Phillip Ferrigon and Kevin Tucker, "Person-First Language vs. Identity-First Language: An Examination of the Gains and Drawbacks of Disability Language in Society," *Journal of Teaching Disability Studies* 1 (January 3, 2019); Krista L. Best et al., "Language Matters! The Long-standing Debate between Identity-first Language and Person First Language," *Assistive Technology* 34, no. 2 (2022): 127–128.

29. Susan Goodman, *Spirit of Stoke Mandeville: The Story of Sir Ludwig Guttmann* (London: HarperCollins, 1986).

30. Bailey, *Athlete First*, 14.

31. Ludwig Guttmann, "Looking Back on a Decade," *The Cord* 6, no. 4 (1954): 9–23.

32. Ludwig Guttmann, "History of the National Spinal Injuries Centre, Stoke Mandeville Hospital, Aylesbury," *Proceeding of the Annual Scientific Meeting of the Society Held at Stoke Mandeville Hospital, Aylesbury*, July 27–29, 1967, 115–126.

33. Rebecca Akkermans, "Ludwig Guttmann," *The Lancet Neurology* 15, no. 12 (2016): 1210.

34. Ludwig Guttmann, *Textbook of Sport for the Disabled* (Aylesbury: HM & M Publishers, 1976), 12.

35. Ludwig Guttmann, "The Second National Stoke Mandeville Games for the Paralyzed," *The Cord* 3 (1949): 24.

36. First used colloquially in a 1953 newspaper article, the term "paralympic" was originally a portmanteau, blending "paraplegic" with "Olympic." It was not until 1988 the IOC approved its official usage. Although the Games have had several names over the years, I use "Paralympic Games" throughout the chapter to avoid confusion.

37. Sean M. Tweedy and Yves C. Vanlandewijck, "International Paralympic Committee Position Stand—Background and Scientific Principles of Classification in Paralympic Sport," *British Journal of Sports Medicine* 45 (2011): 259–269. See also Robert D. Steadward, "Integration and Sport in the Paralympic Movement," *Sport Science Review* 5, no. 1 (1996):

26–41; Yves C. Vanlandewijck and Rudi J. Chappel, "Integration and Classification Issues in Competitive Sports for Athletes with Disabilities," *Sport Science Review* 5, no. 5 (1996): 65–88; Claudine Sherrill, "Disability Sport and Classification Theory: A New Era," *Adapted Physical Activity Quarterly* 16, no. 3 (1999): 206–215; P. David Howe and Carwyn Jones, "Classification of Disabled Athletes:(Dis)empowering the Paralympic Practice Community," *Sociology of Sport Journal* 23, no. 1 (2006): 29–46; David Legg, "Paralympic Games: History and Legacy of a Global Movement," *Physical Medicine Rehabilitation Clinics of North America* 29, no. 2 (2018): 417–425.

38. Bailey, *Athlete First*, 9.

39. Claudine Sherrill, Carol Adams-Mushett, and Jeffrey Al Jones, "Classification and other Issues in Sports for Blind, Cerebral Palsied, Les Autres, and Amuptee Athletes," in *Sport and Disabled Athletes*, ed. Cludine Sherrill (Champaign, IL: Human Kinetics, 1986), 113.

40. Interview with Robert Steadward, July 12, 2021.

41. Ibid.

42. Stan Labanowich and Armand "Tip" Thiboutot, *Wheelchairs Can Jump! A History of Wheelchair Basketball* (Boston: Acanthus, 2011), 189.

43. Ibid.

44. Phillip L. Craven, "The Development from a Medical Classification to a Player Classification in Wheelchair Basketball," *Adapted Physical Activity* (1990): 84.

45. Thiboutot quoted in in Horst Strohkendl, *The 50th Anniversary of Wheelchair Basketball* (New York: Waxman, 1996), 50. On the alienation of classification, see also P. David Howe, *The Cultural Politics of the Paralympic Movement through an Anthropological Lens* (London: Routledge, 2008).

46. Strohkendl, *The 50th Anniversary of Wheelchair Basketball*, 51.

47. Labanowich and Thiboutot, *Wheelchairs Can Jump!* 192.

48. David Legg and Robert Steadward, "The Paralympic Games and 60 Years of Change (1948–2008): Unification and Restructuring from a Disability and Medical Model to Sport-Based Competition," in *Disability in the Global Sport Arena: A Sporting Chance*, ed. Jill M. Le Clair (London: Routledge, 2012), 38.

49. The IPC sanctions three Classification Research and Development Centers: a center for athletes with intellectual impairment at Catholic University of Leuven, Belgium; one for athletes with visual impairments at Vrije Universiteit of Amsterdam, the Netherlands; and the

center devoted to the study of physical impairments at Queensland University, Brisbane, Australia.

50. Interview with Sean Tweedy, October 2, 2019.

51. Sean M. Tweedy, "Taxonomic Theory and the ICF: Foundations for a Unified Disability Athletics Classification," *Adapted Physical Activity Quarterly* 19, no. 2 (2002): 221; 223.

52. IPC Athlete Classification Code.

53. Tweedy and Vanlandewijck, "International Paralympic Committee Position Stand," 265.

54. "IPC Statement on USA Swimmer Victoria Arlen," August 12, 2013, https://www.paralympic.org/news/ipc-statement-usa-swimmer-victoria-arlen.

55. David Wharton, "Paralyzed U.S. Swimmer Banned from Paralympic World Championships," *LA Times,* August 15, 2013, https://www.latimes.com/sports/la-xpm-2013-aug-15-la-sp-sn-paralympic-ban-20130815-story.html.

56. Tweedy, Connick, and Beckman, "Applying Scientific Principles," 322.

57. Interview with Emma Beckman, October 27, 2018.

58. International Paralympic Committee, "What Is Classification," https://www.paralympic.org/classification.

59. International Federation for Athletes with Intellectual Impairments, "Athlete Eligibility," https://inas.org/about-us/athlete-eligibility/eligibility-and-classification. In 2019, the federation changed its name to Virtus.

60. Michael Pavitt, "IPC Says Up to 75 Wheelchair Basketball Athletes Should Undergo Reclassification in Time for Tokyo 2020," *Inside the Games,* February 9, 2020, https://www.insidethegames.biz/articles/1090325/ipc-wheelchair-basketball-tokyo-2020.

61. Quoted in Simon Smale, "New Paralympic Wheelchair Basketball Eligibility Rules Have Ruined Dreams, and Raised Significant Questions," *ABC News,* August 22, 2020, https://www.abc.net.au/news/2020-08-23/paralaympics-wheelchair-basketball-reclassification-ipc/12539302.

62. Sean M. Tweedy, Emma M. Beckman, and Mark J. Connick, "Paralympic Classification: Conceptual Basis, Current Methods, and Research Update," *Physical Medicine and Rehabilitation* 6, no. 8 (2014): S11–S17.

63. International Paralympic Committee, *Models of Best Practice: National Classification* (Bonn: International Paralympic Committee, 2017), https://www.paralympic.org/sites/default/files/document/170216081042262_2017_02_16+Models+of+Best+Practice_National+

Classification.pdf; Mark J. Connick, Emma Beckman, and Sean M. Tweedy, "Evolution and Development of Best Practice in Paralympic Classification," in *The Palgrave Handbook of Paralympic Studies*, ed. Ian Brittain and Aaron Beacom (London: Palgrave Macmillan, 2018), 389–416; Tweedy, Connick, and Beckman, "Applying Scientific Principles," 313–332.

64. Interview with Sean Tweedy, March 16, 2020.

65. Andrew Smith and Nigel Thomas, "The 'Inclusion' of Elite Athletes with Disabilities in the 2002 Manchester Commonwealth Games: An Exploratory Analysis of British Newspaper Coverage," *Sport, Education and Society* 10, no. 1 (2005): 62.

66. Tweedy, Beckman, and Connick, "Paralympic Classification," S13.

67. Interview with Jonna Belanger, September 21, 2020.

68. Ibid.

69. Interview with Shawn Morelli, November 27, 2018. See also Danielle Peers, "Interrogating Disability: The (De)composition of a Recovering Paralympian," *Qualitative Research in Sport, Exercise and Health* 4, no. (2012): 175–188; Bartosz Molik et al., "The International Wheelchair Basketball Federation's Classification System: The Participants' Perspective," *Kinesiology* 49, no. 1 (2017): 117–126; Kirsti Van Dornick and Nancy L. I. Spencer, "What's in A Sport Class? The Classification Experiences of Paraswimmers," *Adapted Physical Activity Quarterly* 37, no. 1 (2020): 1–19.

70. "Hannah Cockroft: 'Para-Classification Tests Are Humiliating,'" October 27, 2017, https://www.bbc.com/sport/disability-sport/41780947.

71. International Paralympic Committee, "Athlete Classification Code," accessed December 22, 2023, https://www.paralympic.org/classificat ion-code.

72. Quoted in "Paralympics: Mallory Weggemann Shocked by Classification Change," *BBC*, August 30, 2012, https://www.bbc.com/sport/disabil ity-sport/19429915.

73. "World Para Athletics Announces Classification Changes," *Athletics Weekly*, October 26, 2017, https://athleticsweekly.com/athletics-news/ world-para-athletics-announces-classification-changes-69600/.

74. International Paralympic Committee, *IPC Classification Code: Models of Best Practice, Intentional Misrepresentation Rules* (Bonn: International Paralympic Committee, 2013), https://m.paralympic.org/sites/defa ult/files/document/141113161802225_2014_10_10+Sec+ii+chapter+ 1_3_Models+of+best+practice_+Intentional+Misrepresentation+ Rules.pdf.

75. Tracey Holmes, "Allegations of Cheating, Threats and Cover-ups Aimed at Australian Paralympic Swimming," *ABC News,* December 2, 2017, http://www.abc.net.au/news/2017-12-03/allegations-target-aus tralian-paralympic-swimming/9221084.

76. Knapp, "Why Ukraine's Small Paralympic Team Packs Such a Big Punch."

77. Deidra Dionne, "Are Some Countries Hacking the Paralympic System to Win More Medals?" November 15, 2016, https://www.cbc.ca/spo rts/olympics/2.6605/paralympics-classification-system-1.3850121.

78. Quoted in Dan Barnes, "'I'm Number One in the World—Other than Ukraine': Eastern Euro Trio Thwarts Legendary Canadian Swimmer," *National Post,* September 12, 2016, https://nationalpost.com/sports/ olympics/im-number-one-in-the-world-other-than-ukraine-eastern- european-trio-thwarts-legendary-canadian-swimmer.

79. Silverman, "Ian Silverman."

80. International Paralympic Committee, *Position Statement Regarding the Participation of Athletes with an Intellectual Disability at IPC Sanctioned Events* (Bonn: International Paralympic Committee, 2016).

81. Steadman interview.

82. Holmes, "Allegations of Cheating."

83. UK Sport, "UK Sport Statement on Funding," November 25, 2004, https://www.uksport.gov.uk/news/2004/11/25/uk-sport-statement- on-funding

84. Quoted in Paul Grant, "'I'm Handing Back My Medal': Is Paralympic Sport Classification Fit for Purpose?" *BBC,* September 18, 2017, http:// www.bbc.com/sport/disability-sport/41253174. See also Andy Brown, "Paralympic Classification System Allegedly Open to Abuse," *Sport Integrity Initiative,* November 2, 2017, https://www.sportsintegrityinitiat ive.com/paralympic-classification-system-allegedly-open-abuse/.

85. Mike Kelly, "Paralympic Cheats Have Cost Me Medals Says Cramlington Legend Stephen Miller," *Chronicle,* November 7, 2017, https://www. chroniclelive.co.uk/news/north-east-news/paralympic-cheats-cost- medals-says-13864613.

86. World Para Winter Sports, "The IPC Provide an Update on Alleged Cases of Intentional Misrepresentation," August 11, 2016, https://www. paralympic.org/news/ipc-provide-update-alleged-cases-intentional- misrepresentation.

87. Leonardo José Mataruna-Dos-Santos, Andressa Fontes Guimarães- Mataruna, and Daniel Range, "Paralympians Competing in the Olympic Games and the Potential Implications for the Paralympic

Games," *Brazilian Journal of Education, Technology, and Society* 11, no. 1 (2018): 105–116.

88. Most accounts describe Eyser's prosthesis as wooden, but at least one contemporary account held that Eyser "wears an artificial leg when running and jumping. When exercising on the horizontal bar or the "horse," however, he either removes the cork leg or straps it closely to his sound limb." "One-Legged Man an Athlete," *Morning Oregonian,* May 29, 1908, 7.

89. International Association of Athletics Federations, *IAAF Competition Rules,* 2008, https://www.worldathletics.org/404?aspxerrorpath=/newsfiles/42192.pdf.

90. Court of Arbitration for Sport, *Arbitration CAS 2008/A/1480 Pistorius v/ IAAF, award of 16 May 2008,* https://jurisprudence.tas-cas.org/Shared%20Documents/1480.pdf.

91. See International Association of Athletics Federations, "Oscar Pistorius—Independent Scientific Study Concludes That Cheetah Prosthetics Offer Clear Mechanical Advantages," January 14, 2008, https://www.worldathletics.org/news/news/oscar-pistorius-independent-scientific-stud-1; Peter G. Weyand et al., "The Fastest Runner on Artificial Legs: Different Limbs, Similar Function?" *Journal of Applied Physiology* 107, no. 3 (2009): 903–911.

92. Emphasis added. The rule change resulted from the exceptional long-jump performance of Markus Rehm, who uses one prosthetic limb. Rehm's Para world record stands at 8.62 meters, which falls short of Mike Powell's 1991 world record of 8.95 meters but would have won Rehm gold at the 2004, 2008, 2012, 2016, and 2020 Olympic Games.

93. Court of Arbitration for Sport, CAS 2020/A/6807 Blake Leeper v. International Association of Athletics Federations, accessed December 22, 2023, https://www.tas-cas.org/fileadmin/user_upload/Award__6807___for_publication_.pdf.

94. Ibid.

95. Ibid.

96. Once again, researchers disagreed about Leeper's advantage. See Paolo Taboga, Owen N. Beck, and Alena M. Grabowski, "Prosthetic Shape, but Not Stiffness or Height, Affects the Maximum Speed of Sprinters with Bilateral Transtibial Amputations," *PloS One* 15, no. 2 (2020): e0229035–e0229035; Peter G. Weyand et al., "Artificially Long Legs Directly Enhance Long Sprint Running Performance," *Royal Society Open Science* 9, no. 8 (2022): 220397.

97. Court of Arbitration for Sport, "Media Release: The Court of Arbitration for Sport (CAS) Partially Upholds the Appeal of Blake Lepper," accessed December 22, 2023, https://www.tas-cas.org/filead min/user_upload/CAS_Media_Release_6807.pdf.

98. Emma M. Beckman et al., "Should Markus Rehm Be Permitted to Compete in the Long Jump at the Olympic Games?" *British Journal of Sports Medicine* 51, no. 14 (2017): 1048–1049.

99. Ibid.

100. Anne Marcellini et al., "Challenging Human and Sporting Boundaries: The Case of Oscar Pistorius," *Performance Enhancement & Health* 1, no. 1 (2012): 3–9.

CHAPTER 4

1. Willy Voet, *Breaking the Chain, Drugs and Cycling: The True Story*, trans. William Fotheringham (London: Yellow Jersey Press, 2001).

2. Christophe Brissonneau, "The 1998 Tour de France: Festina, from Scandal to an Affair in Cycling," in *Routledge Handbook of Drugs and Sport*, ed. Verner Møller, Ivan Waddington, and John M. Hoberman (London: Routledge, 2015), 181–192.

3. Dag Vidar Hanstad, Andy Smith, and Ivan Waddington, "The Establishment of the World Anti-Doping Agency: A Study of the Management of Organizational Change and Unplanned Outcomes," *International Review for the Sociology of Sport* 43, no. 3 (September 2008): 227–249.

4. John Hoberman, "How Drug Testing Fails: The Politics of Doping Control," in *Doping in Elite Sport: Politics of Drugs in the Olympic Movement*, ed. Wayne Wilson and Edward Derse (Champaign, IL: Human Kinetics, 2001), 242.

5. McCaffrey and Banks quoted in Duncan Mackay, "Tony Banks Criticises IOC at the World Conference on Doping," *The Guardian*, February 3, 1999, https://www.theguardian.com/sport/1999/feb/03/tony-banks-criticises-ioc-conference-doping-sport.

6. Maarten van Bottenburg, Arnout Geeraert, and Oliver de Hon, "The World Anti-Doping Agency: Guardian of Elite Sport's Credibility," in *Guardians of Public Value: How Public Organizations Become and Remain Institutions*, ed. Arjen Boin, Lauren A. Fahy, and Paul 't Hart (Cham, Switzerland: Palgrave Macmillan, 2021), 193.

7. World Anti-Doping Agency, World Anti-Doping Code, accessed December 22, 2023, https://www.wada-ama.org/en/resources/world-anti-doping-program/world-anti-doping-code.

8. Ivan Waddington and Verner Møller, "WADA at Twenty: Old Problems and Old Thinking?" *International Journal of Sport Policy and Politics* (2019), doi: 10.1080/19406940.2019.1581645.

9. World Anti-Doping Agency, *2020 Anti-Doping Testing Figures Report*, accessed December 22, 2023, https://www.wada-ama.org/sites/default/files/2022-01/2020_anti-doping_testing_figures_en.pdf.

10. Olivier de Hon, Harm Kuipers, and Maarten van Bottenburg, "Prevalence of Doping Use in Elite Sports: A Review of Numbers and Methods," *Sports Medicine*, 45, no. 1 (2015): 57–69.

11. Umair Irgan and Julia Bellus, "Olympic Swimmer Ryan Lochte Broke Doping Rules. It Happens Far More Than You Think," *Vox,* July 27, 2018, https://www.vox.com/2018/7/24/17603358/ryan-lochte-doping-ban-olympics-instagram.

12. Rolf Ulrich et al., "Doping in Two Elite Athletics Competitions Assessed by Randomized-Response Surveys," *Sports Medicine* 48, no. 1 (2018): 211–219.

13. Ludwig Prokop, "The Struggle Against Doping and Its History," *Journal of Sports Medicine and Physical Fitness* 10, no. 1 (1970): 45–48.

14. Charles E. Yesalis and Michael S. Bahrke, "History of Doping in Sport," *International Sports Studies* 24, no. 1 (2002): 42–76.

15. Personal correspondence, August 23, 2018.

16. "'Dope' Evil of the Turf," *New York Times*, October 19, 1903, 8.

17. Wray Vamplew, "Playing with the Rules: Influences on the Development of Regulation in Sport," *International Journal of the History of Sport* 24, no. 7 (2007): 843–871.

18. John Gleaves, "Enhancing the Odds: Horse Racing, Gambling and the First Anti-Doping Movement in Sport, 1889–1911," *Sport in History* 32, no. 1 (2012): 26–52.

19. Charles J. P. Lucas, *The Olympic Games, 1904* (St. Louis, MO: Woodward and Tiernan, 1905), 51.

20. John Gleaves, "Doped Professionals and Clean Amateurs: Amateurism's Influence on the Modern Philosophy of Anti-Doping," *Journal of Sport History* 38, no. 2 (2011): 237–254. On doping in other sports during the 1920s, see Neil Carter, "Monkey Glands and the Major: Frank Buckley and Modern Football Management," Manchester Metropolitan University, 2011, https://dora.dmu.ac.uk/bitstream/handle/2086/4617/Frank%20Buckley%20Essay.pdf?sequence=3; John Hoberman,

Mortal Engines: The Science of Performance and the Dehumanization of Sport (New York: Free Press, 1992), 142.

21. Jörg Krieger, Lindsay Parks Pieper, and Ian Ritchie, "Sex, Drugs and Science: The IOC's and IAAF's Attempts to Control Fairness in Sport," *Sport in Society* 22, no. 9 (2019): 1555–1573.

22. International Amateur Athletic Federation, *Handbook of the International Amateur Athletic Federation, 1927–1928,* accessed December 22, 2023, https://www.iaaf.org/news/news/a-piece-of-anti-doping-history-iaaf-handbook. Emphasis added.

23. International Olympic Committee, *Bulletin Officiel Du Comite International Olympique* no. 37 (July 1938), quoted in Gleaves, "Doped Professionals," 247. Emphasis added.

24. Ove Bøje, "Doping: A Study of the Means Employed to Raise the Level of Performance in Sport," *Bulletin of the Health Organization of the League of Nations* 8 (1939): 439.

25. Nicolas Rasmussen, *On Speed: From Benzedrine to Adderall* (New York: NYU Press, 2008).

26. Louis Lasagna, "The Pharmaceutical Revolution: Its Impact on Science and Society," *Science* 166, no. 3910 (1969): 1227–1233.

27. Verner Møller, "Knud Enemark Jensen's Death During the 1960 Rome Olympics: A Search for Truth?," *Sport in History* 25, no. 3 (2005): 452–471.

28. Christopher S. Thompson, *The Tour de France: A Cultural History* (Berkeley: University of California Press, 2006), 232–233. Athletes at the 1966 FIFA World Cup also underwent testing.

29. William Fotheringham, *Put Me Back on my Bike: In Search of Tom Simpson* (New York: Random House, 2012).

30. Bernat López, "Creating Fear: The 'Doping Deaths,' Risk Communication and the Anti-Doping Campaign," *International Journal of Sport Policy and Politics* 6, no. 2 (2014): 213–225; Patrick Mignon, "The Tour de France and the Doping Issue," *International Journal of the History of Sport* 20, no. 2 (2003): 231.

31. "Medical Commission," *Olympic Newsletter* 5 (1968): 71–73.

32. Quoted in Paul Dimeo, *A History of Drug Use in Sport 1876–1976: Beyond Good and Evil* (London: Routledge, 2008), 114.

33. Don H. Catlin, Kenneth D. Fitch, and Arne Ljungqvist, "Medicine and Science in the Fight Against Doping in Sport," *Journal of Internal Medicine* 264, no. 2 (2008): 101.

34. Terry Todd, "Anabolic Steroids: The Gremlins of Sport," *Journal of Sport History* 14, no. 1 (1987): 87–107.

35. Jan Todd and Terry Todd, "Significant Events in the History of Drug Resting and the Olympic Movement: 1960–1999," in *Doping in Elite Sport: Politics of Drugs in the Olympic Movement*, ed. Wayne Wilson and Edard Derse (Champaign, IL: Human Kinetics, 2001), 73.

36. Paul Dimeo, Thomas M. Hunt, and Richard Horbury, "The Individual and the State: A Social Historical Analysis of the East German 'Doping System,'" *Sport in History* 31, no. 2 (2011): 218–237; Steven Ungerleider and Bill Bradley, *Faust's Gold: Inside the East German Doping Machine* (New York: Macmillan, 2001).

37. Werner W. Franke and Brigitte Berendonk, "Hormonal Doping and Androgenization of Athletes: A Secret Program of the German Democratic Republic Government," *Clinical Chemistry* 43, no. 7 (1997): 1262–1279.

38. Mike Dennis, "Securing the Sports 'Miracle': The Stasi and East German Elite Sport," *International Journal of the History of Sport* 29, no. 18 (2012): 2554.

39. Mike Dennis, "The East German Doping Programme," in *Routledge Handbook of Drugs in Sport*, 177. See also Thomas M. Hunt et al., "The Health Risks of Doping during the Cold War: A Comparative Analysis of the Two Sides of the Iron Curtain," *International Journal of the History of Sport* 31, no. 17 (2014): 2230–2244.

40. Grace Huang and Shehzad Basaria, "Do Anabolic-androgenic Steroids Have Performance-Enhancing Effects in Female Athletes?," *Molecular and Cellular Endocrinology* 464 (2018): 56–64.

41. Daniel Read et al., *WADA, the World Anti-Doping Agency: A Multi-Level Legitimacy Analysis* (London: Routledge, 2021), 3. Suspicions of state-sponsored doping programs have been leveled at the Soviet Union, China, Bulgaria, and others. See Bruce Kidd, Robert Edelman, and Susan Brownell, "Comparative Analysis of Doping Scandals: Canada, Russia, and China," in *Doping in Elite Sport: The Politics of Drugs in the Olympic Movement*, ed. Wayne Wilson and Ed Derse (Champaign, IL: Human Kinetics, 2001), 153–188; Yesalis and Bahrke, "History of Doping in Sport."

42. Björn Ekblom, Alberto N. Goldbarg, and Bengt Gullbring, "Response to Exercise After Blood Loss and Reinfusion," *Journal of Applied Physiology* 33, no. 2 (1972): 175–180. By the late 1980s and 1990s, athletes achieved similar results with recombinant EPO, a synthetic version of a natural hormone originally developed to treat anemia.

43. Quoted in Robert Sullivan, "Triumphs Tainted with Blood," *Sports Illustrated,* January 21, 1985, https://www.si.com/vault/1985/01/

21/546256/triumphs-tainted-with-blood. See also John Gleaves, "Manufactured Dope: How the 1984 US Olympic Cycling Team Rewrote the Rules on Drugs and Sports," *International Journal of the History of Sport* 32, no. 1 (2015): 89–107.

44. Alister Browne, Victor Lachance, and Andrew Pipe, "The Ethics of Blood Testing as an Element of Doping Control in Sport," *Medicine and Science in Sports and Exercise* 31, no. 4 (1999): 497–501.

45. Duncan Mackay, "The Dirtiest Race in History: Olympic 100m Final, 1988," *The Guardian,* April 17, 2003, https://www.theguardian.com/ sport/2003/apr/18/athletics.comment. Chris E. Cooper argues that the women's 1500-meter race was actually "dirtier" than the 1988 men's 100m final. In *Run, Swim, Throw, Cheat: The Science Behind Drugs in Sport* (New York: Oxford University Press, 2012), 11–13.

46. Michael Janofsky and Peter Alfano, "Drug Use by Athlete Runs Free Despite Tests," *New York Times,* November 17, 1988, https://www.nyti mes.com/1988/11/17/sports/drug-use-by-athletes-runs-free-despite-tests.html.

47. Thomas M. Hunt, *Drug Games: The International Olympic Committee and the Politics of Doping, 1960–2008* (Austin: University of Texas Press, 2011).

48. Norway first utilized out-of-competition in the 1970s, as did Sweden the following decade.

49. World Anti-Doping Agency, "Frequently Asked Questions," accessed December 22, 2023, https://www.wada-ama.org/sites/default/files/ resources/files/athlete_central_faq_final_en_0.pdf.

50. World Anti-Doping Agency, Athlete Whereabouts At-a-Glance, accessed December 22, 2023, https://www.wada-ama.org/sites/default/ files/2022-03/At_a_Glance_Whereabouts_English_Live_2021.pdf.

51. Ivan Waddington, "'A Prison of Measured Time'? A Sociologist Looks at the WADA Whereabouts System," in *Doping and Anti-Doping Policy in Sport: Ethical, Legal and Social Perspectives*, ed. Mike McNamee et al. (London: Taylor & Francis, 2011), 186. See also Verner Møller, "One Step Too Far—About WADA's Whereabouts Rule," *International Journal of Sport Policy and Politics* 3, no. 2 (2011): 178.

52. Nadal quoted in "Testers Treat Us Like Criminals, Says Nadal," *The Guardian,* February 12, 2009, https://www.theguardian.com/sport/ 2009/feb/12/rafael-nadal-drugs-test-criminals-olympic-andy-murray.

53. Paul Dimeo and Verner Møller, *The Anti-Doping Crisis in Sport: Causes, Consequences, Solutions* (London: Routledge, 2018).

54. Dag Vidar Hanstad and Sigmund Loland, "Elite Athletes' Duty to Provide Information on Their Whereabouts: Justifiable Anti-Doping

Work or an Indefensible Surveillance Regime?" *European Journal of Sport Science* 9, no. 1 (2009): 3–10.

55. Quoted in Joseph Doherty, "From Ben Johnson to Barry Bonds: Sports' Steroid Scandal Continues," *Bleacher Report,* April 18, 2010, https://bleacherreport.com/articles/381072-from-ben-johnson-to-barrybonds-sports-steroid-scandal-continues.

56. Angela J. Schneider, "Privacy, Confidentiality and Human Rights in Sport," *Sport in Society* 7, no. 3 (2004): 452.

57. Mark Fainaru-Wada and Lance Williams, *Game of Shadows: Barry Bonds, BALCO, and the Steroids Scandal That Rocked Professional Sports* (New York: Penguin, 2006).

58. See John Gleaves, "Biometrics and Antidoping Enforcement in Professional Sport," *American Journal of Bioethics* 17, no. 1 (2017): 77–79.

59. WADA, World Anti-Doping Code, International Standard, Prohibited List, 2024, accessed December 22, 2023, https://www.wada-ama.org/sites/default/files/2023-09/2024list_en_final_22_september_2023.pdf.

60. WADA, World Anti-Doping Code, 33–34.

61. WADA, World Anti-Doping Code, 34.

62. Jules A. A. C. Heuberger and Adam F. Cohen, "Review of WADA Prohibited Substances: Limited Evidence for Performance-Enhancing Effects," *Sports Medicine* 49, no. 4 (2019): 525–539; Eduard Bezuglov et al., "The Inclusion in WADA Prohibited List is Not Always Supported by Scientific Evidence: A Narrative Review," *Asian Journal of Sports Medicine* 12, no. 2 (2021).

63. Harrison G. Pope Jr. et al., "Adverse Health Consequences of Performance-Enhancing Drugs: An Endocrine Society Scientific Statement," *Endocrine Reviews* 35, no. 3 (2014): 341–375.

64. David van Mill, "Why Are We So Opposed to Performance-Enhancing Drugs in Sport?" *The Conversation,* August 27, 2015, http://theconversation.com/why-are-we-so-opposed-to-performance-enhancing-drugs-in-sport-46528.

65. World Anti-Doping Agency, "WADA Note on Artificially Induced Hypoxic Conditions," 2006, http://www.mcst.go.kr/servlets/eduport/front/upload/UplDownloadFile?pFileName=Note%20 on%20Hypoxia%20May%202006.pdf&pRealName = F362307.pdf&pPath = 040415000.

66. Torbjörn Tännsjö, "Medical Enhancement and the Ethos of Elite Sport," in *Human Enhancement*, ed. Julian Savulescu and Nick Bostrom (New York: Oxford University Press), 316.

67. Leon R. Kass, *Beyond Therapy: Biotechnology and the Pursuit of Happiness, A Report of The President's Council on Bioethics* (New York: Dana Press, 2003), 4; Hoberman, *Mortal Engines*.

68. Bengt Kayser, Alexandre Mauron, and Andy Miah, "Current Ant-Doping Policy: A Critical Appraisal," *BMC Medical Ethics* 8, no. 2 (2007), https://bmcmedethics.biomedcentral.com/articles/10.1186/1472-6939-8-2.

69. Thomas Douglas, "Enhancement in Sport, and Enhancement Outside Sport," *Studies in Ethics, Law, and Technology* 1, no. 1 (2009): ukpmcpa2293.

70. J. G. P. Williams, *Sports Medicine* (London: Edward Arnold, 1962), vii.

71. William Saletan, "The Beam in Your Eye: If Steroids Are Cheating, Why Isn't LASIK?" *Slate,* April 18, 2005, https://slate.com/technology/2005/04/if-steroids-are-cheating-why-isn-t-lasik.html.

72. J. David Kopp, "Eye on the Ball: An Interview with Dr. C. Stephen Johnson and Mark McGwire," *Journal of the American Optometric Association* 70, no. 2 (1999): 79–84.

73. Tännsjö, "Medical Enhancement," 320–321.

74. Mark Bailey, "Miguel Indurain: The Record Tour Winner," *Cyclist,* May 31, 2016, https://www.cyclist.co.uk/in-depth/423/miguel-indurain-the-record-tour-winner.

75. D. W. Lawrence, "Sociodemographic Profile of an Olympic Team," *Public Health* 148 (2017): 149–158.

76. Andrew Novak, "Disability Sport in Sub-Saharan Africa: From Economic Underdevelopment to Uneven Empowerment," *Disability and the Global South* 1, no. 1 (2014): 44–63.

77. WADA Ethics Panel: Guiding Values in Sport and Anti-Doping, 2017, https://www.wada-ama.org/sites/default/files/resources/files/wada_ethicspanel_setofnorms_oct2017_en.pdf

78. WADA, World Anti-Doping Code, 34. Emphasis in the original.

79. World Anti-Doping Agency, *Therapeutic Use Exemptions* (Montreal: World Anti-Doping Agency, 2021), https://www.wada-ama.org/sites/default/files/resources/files/international_standard_istue_-_2021.pdf.

80. Yannis Pitsiladis et al., "Make Sport Great Again: The Use and Abuse of the Therapeutic Use Exemptions Process," *Current Sports Medicine Reports* 16, no. 3 (2017): 123–125.

81. Kayser, Mauron, and Miah, "Current Ant-Doping Policy."

82. Ivan Waddington, "Theorising Unintended Consequences of Anti-Doping Policy," *Performance Enhancement & Health* 4, no. 3–4 (2016): 80–87.

83. Philip B. Maffetone and Paul B. Laursen, "Athletes: Fit but Unhealthy?" *Sports Medicine—Open* 2, no. 1 (2016), 24.

84. Kevin Young, "Violence, Risk, and Liability in Male Sports Culture," *Sociology of Sport Journal* 10, no. 4 (1993): 373–396.

85. Sally Jenkins, "There's a Legal Remedy to the Doping Issue," *Washington Post,* October 12, 2007, E1; Abott quoted in Nora Caplan-Bricer, "The Inextricable Tie Between Eating Disorders and Endurance Athletes," *Outside,* June 23, 2017, https://www.outsideonline.com/2191906/eating-disorders-are-more-common-you-think#:~:text=%E2%80%9CBeing%20a%20professional%20athlete%20isn,much%20a%20body%20can%20withstand.%E2%80%9D.

86. Norman Frost, "Banning Drugs in Sports: A Skeptical View," *Hastings Center Report* 16, no. 4 (1986): 9.

87. Ian Ritchie, "The Construction of a Policy: The World Anti-Doping Code's 'Spirit of Sport' Clause," *Performance Enhancement & Health* 2, no. 4 (2013): 197; Bennett Foddy and Julian Savulescu, "Ethics of Performance Enhancement in Sport: Drugs and Gene Doping," *Principles of Health Care Ethics* (2006): 511–519; Mojisola Obasa and Pascal Borry, "The Landscape of the 'Spirit of Sport,'" *Journal of Bioethical Inquiry* 16, no. 3 (2019): 443–453; Michael J. McNamee, "The Spirit of Sport and the Medicalisation of Anti-Doping: Empirical and Normative Ethics," *Asian Bioethics Review* 4, no. 4 (2012): 374–392.

88. WADA, World Anti-Doping Code, 13.

89. Michael J. McNamee, "The Spirit of Sport and Anti-Doping Policy: An Ideal Worth Fighting For," *Play True* 1 (2013): 14–16. See also Sigmund Loland and Michael J. McNamee, "The 'Spirit of Sport,' WADAs Code Review, and the Search for An Overlapping Consensus," *International Journal of Sport Policy and Politics* 11, no. 2 (2019): 325–339.

90. Claudio Tamburrini, "Are Doping Sanctions Justified? A Moral Relativistic View," *Sport in Society* 9, no. 2 (2006): 200, 203.

91. Heikki R. Rusko, "New Aspects of Altitude Training," *American Journal of Sports Medicine* 24, no. 6 (1996): S48–S52; Randall L. Wilber, "Application of Altitude/Hypoxic Training by Elite Athlete," *Journal of Human Sport and Exercise* 6, no. 2 (2011): 273.

92. David J. Armstrong, "Hypoxic Chambers and Other Artificial Environments," *Drugs in Sport*, ed. Davide R. Mottram (London: Routledge, 2012), 354–369.

93. For counterargument, see Giuseppe Lippi, Massimo Franchini, and Gian Cesare Guidi, "Prohibition of Artificial Hypoxic Environments in Sports: Health Risks Rather Than Ethics," *Applied Physiology, Nutrition, and Metabolism* 32, no. 6 (2007): 1206–1207.

94. Quoted in Doriane Lambelet Coleman et al., "Position Paper of the Center for Sports Law and Policy: Whether Artificially Induced Hypoxic Conditions Violate the 'Spirit of Sport,'" Center for Sports Law & Policy, Duke Law, 2006, https://law.duke.edu/features/pdf/hypoxiaresponse.pdf.

95. Quoted in Benjamin D. Levine, "Should 'Artificial' High Altitude Environments Be Considered Doping?" *Scandinavian Journal of Medicine and Science in Sports* 16 (2006): 299.

96. Andy Miah, "Rethinking Enhancement in Sport," *Annals New York Academy of Sciences,* 1093, no. 1 (2006): 301–320.

97. World Anti-Doping Agency, *WADA Note on Artificially Induced Hypoxic Conditions* (2006), quoted in Eric T. Juengst, "Subhuman, Superhuman, and Inhuman: Human Nature and the Enhanced Athlete," in *Athletic Enhancement, Human Nature, and Ethics,* ed. Jan Tolleneer, Sigrid Sterckx, and Pieter Bonte (Dordrecht: Springer, 2013), 97.

98. Levine, "Should 'Artificial,'" 299.

99. Quoted in "Ani-Doping Agency Won't Ban Oxygen Tents," *CBC Sports,* September 16, 2006, https://www.cbc.ca/sports/anti-doping-agency-won-t-ban-oxygen-tents-1.589665.

100. Angela Schneider, "Banned from the Tokyo Olympics for Pot? Let the Athletes Decide What Drugs Should Be Allowed," *The Conversation,* July 6, 2021, https://theconversation.com/banned-from-the-tokyo-olympics-for-pot-let-the-athletes-decide-what-drugs-should-be-allowed-163619.

101. Ellis Cashmore, "Opinion: It's Time to Allow Doping in Sport," *CNN,* October 24, 2012, https://www.cnn.com/2012/10/23/opinion/cashmore-time-to-allow-doping-in-sport/index.html.

102. Thomas Douglas, "Enhancement in Sport, and Enhancement Outside Sport," *Studies in Ethics, Law, and Technology* 1, no. 1 (2009): ukpmcpa2293.

103. Dimeo Møller, *The Anti-Doping Crisis in Sport.*

104. Michael Veber, "The Coercion Argument Against Performance-Enhancing Drugs," *Journal of the Philosophy of Sport* 41, no. 2 (2014): 267–277; David A. Baron, David M. Martin, and Samir Abol Magd, "Doping in Sports and Its Spread to At-Risk Populations: An International Review," *World Psychiatry* (2007): 118–123.

105. Aaron C. T. Smith and Bob Stewart, "Why the War on Drugs in Sport Will Never Be Won," *Harm Reduction Journal* 12, no. 51 (2015): 53

106. Robin Parisotto, *Blood Sports: The Inside Dope on Drugs in Sport* (Melbourne: Hardie Grant, 2006), location 500.

107. International Olympic Committee, Annual Report, 2021, https://
stillmed.olympics.com/media/Documents/International-Olympic-
Committee/Annual-report/IOC-Annual-Report-2021.pdf?_ga=
2.233961374.1231423006.1662651976-399267994.1662651976.

108. Barrie Houlihan et al., "The World Anti-Doping Agency at 20: Progress
and Challenges," *International Journal of Sport Policy and Politics* 11, no. 2
(2019): 193–201.

109. Carsten Kraushaar Martensen and Verner Møller, "More Money—
Better Anti-Doping?" *Drugs: Education, Prevention and Policy* 24, no. 3
(2017): 286–294.

110. "CIRC Report: Executive Summary," March 2015, https://www.velon
ews.com/2015/03/news/circ-report-executive-summary_362351 .

111. Kayser, Mauron, and Miah, "Current Ant-Doping Policy."

112. Claudio M. Tamburrini, "What's Wrong with Doping?," in *Values in
Sport: Elitism, Nationalism, Gender Equality and the Scientific Manufacture
of Winners*, ed. Torbjörn Tännsjö and Claudio Tamburrini (London:
E&FN Spon, 2000), 201; Karen J. Maschke, "Performance-Enhancing
Technologies and the Ethics of Human Subjects Research," in
*Performance-Enhancing Technologies in Sports: Ethical, Conceptual, and
Scientific Issues*, ed. Thomas H. Murray, Karen J. Maschke, and Angela A.
Wasunna (Baltimore: Johns Hopkins University Press, 2009), 98.

113. Eric Moore and Jo Morrison, "In Defense of Medically Supervised
Doping," *Journal of the Philosophy of Sport* (2022): 167.

114. Bengt Kayser and Jan Tolleneer, "Ethics of a Relaxed Antidoping Rule
Accompanied by Harm-Reduction Measures," *Journal of Medical Ethics*
43 (2017): 283. See also Bengt Kayser and Barbara Broers, "Doping and
Performance Enhancement: Harms and Harm Reduction," in *Routledge
Handbook of Drugs and Sport*, ed. Verner Møller, Ivan Waddington, and
John Hoberman (London: Routledge, 2015), 363–376; Ken Kirkwood,
"Considering Harm Reduction as the Future of Doping Control
Policy in International Sport," *Quest* 61, no. 2 (2009): 180–190.

115. Julian Savulescu, "Why It's Time to Legalize Doping in Athletics," *The
Conversation*, August 28, 2015, https://theconversation.com/why-its-
time-to-legalise-doping-in-athletics-46514.

116. Mike McNamee, "Doping Scandals, Rio, and the Future of Anti-
Doping Ethics," *Sport, Ethics and Philosophy* 10, no. 2 (2016): 115. See also
Søren Holm, "Doping under Medical Control—Conceptually Possible
but Impossible in the World of Professional Sports?," *Sports, Ethics and
Philosophy* 1, no. 2 (2007): 135–145.

117. Jan Todd and Terry Todd, "Reflections on the 'Parallel Federation Solution' to the Problem of Drug Use in Sport," in *Performance-Enhancing Technologies in Sports: Ethical, Conceptual, and Scientific Issues*, ed. Thomas H. Murray, Karen J. Maschke, and Angela A. Wasunna (Baltimore: Johns Hopkins University Press, 2009), 46.

CHAPTER 5

1. Marlen Garcia, "Attorney: Curry Won't Take DNA Test," *Chicago Tribune*, September 24, 2005, sec. 3, p. 1.
2. Quoted in Alan Schwarz, "Baseball's Use of DNA Tests Raises Ethical Issues," *International Herald Tribune*, July 22, 2009, http://www.nytimes.com/2009/07/22/sports/baseball/22dna.html.
3. "The Inside Story of a Toxic Culture at Maryland Football," *ESPN*, August 10, 2018, https://www.espn.com/college-football/story/_/id/24342005/maryland-terrapins-football-culture-toxic-coach-dj-durkin.
4. Iliana Limón, "Sickle Cell Trait: The Silent Killer," *Orlando Sentinel*, July 24, 2011, C1.
5. Quoted in Moisekapenda Bower, "Parents of Former Rice DB File Wrongful Death Suit," *Chronicle*, August 9, 2011, https://www.chron.com/sports/college-football/article/Parents-of-former-Rice-DB-file-wrongful-death-suit-1788862.php.
6. The sickle cell trait policy was initially just for Division I athletes, but the NCAA extended it to cover Division II in 2012, and Division III athletes in 2013. A student-athlete must provide documented results from a previous sickle cell solubility test or undergo testing during their preparticipation medical examination. See Mary G. McDonald, "Screening Saviors? The Politics of Care, College Sports, and Screening Athletes for Sickle Cell Trait," in *Sports, Society, and Technology: Bodies, Practices, and Knowledge Production*, ed. Jennifer J. Sterling and Mary G. McDonald (Gate East, Singapore: Palgrave Macmillan, 2020), 247–267.
7. Benjamin K. Buchanan et al., "Sudden Deaths Associated with Sickle Cell Trait Before and After Mandatory Screening," *Sports Health: A Multidisciplinary Approach* 12, no. 3 (2020), https://doi.org/10.1177/1941738120915690.
8. As per collective bargaining agreements, the NFL also screens for sickle cell trait, as well as Tay Sachs disease (caused by variants in the *HEXA* gene), and mutations of the *G6PD* gene associated with anemia.

9. Elina Rantanen et al., "What Is Ideal Genetic Counselling? A Survey of Current International Guidelines," *European Journal of Human Genetics* 16, no. 4 (2008): 445–452.

10. Janis L. Abkowitz, "President's Column—Sickle Cell Trait and Sports: Is the NCAA a Hematologist?" *The Hematologist* 10, no. 3 (2013), https://doi.org/10.1182/hem.V10.3.1070

11. Quoted in "'Junction Boys Syndrome: How College Football Fatalities Became Normalized," *The Guardian*, August 19, 2018, https://www.theguardian.com/sport/2018/aug/19/college-football-deaths-offseason-workouts.

12. Abkowitz, "President's Column"; Kimberly G. Harmon et al., "Sickle Cell Trait Associated with a RR of Death 37 times in National Collegiate Athletic Association Football Athletes: A Database with 2 Million Athlete-Years as the Denominator," *British Journal of Sports Medicine* 46, no. 5 (2012): 325–330.

13. "CRISPR-Cas9 Gene Editing for Sickle Cell Disease and β-Thalassemia," *New England Journal of Medicine* 384 (2021): 252–260.

14. Owen Slot, "Apocalypse Now: Fears of Gene Doping Are Realised," *The Times,* February 2, 2006, 78.

15. Elaine A. Ostrander, Heather J. Hudson, and Gary K. Ostrander, "Genetics of Athletic Performance," *Annual Review of Genomics and Human Genetics* 10 (2009): 407–429; Molly S. Bray et al., "The Human Gene Map Performance and Health-Related Fitness Phenotypes: The 2006–2007 Update," *Medicine and Science in Sports and Exercise* 41, no. 1 (2009): 35–73.

16. Personal correspondence, July 2, 2018.

17. Hugh E. Montgomery et al., "Human Gene for Physical Performance," *Nature* 393 (May 21, 1998): 221–222; George Gayagay et al., "Elite Endurance Athletes and the ACE I Allele: The Role of Genes in Athletic Performance," *Human Genetics* 103, no. 1 (1998): 48–50.

18. Fabio Andre Castilha et al., "The Influence of Gene Polymorphisms and Genetic Markers in the Modulation of Sports Performance: A Review," *Journal of Exercise Physiology Online* 21, no. 2 (2018): 248–265.

19. Nan Yang et al., "ACTN3 Genotype is Associated with Human Elite Athletic Performance," *American Journal of Human Genetics* 73, no. 3 (2003): 627–631.

20. Ibid. For contradictory findings, see E. D. Hanson et al., "ACTN3 Genotype Does Not Influence Muscle Power," *International Journal of Sports Medicine* 31, no. 11 (2010): 834–838; Myosotis Massidda et al., "ACTN3 R577X Polymorphism Is Not Associated with Team Sport

Athletic Status in Italians," *Sports Medicine* 1, no. 1 (2015): 1–5; Ioannis D. Papadimitriou et al., "No Association Between ACTN3 R577X and ACE I/D Polymorphisms and Endurance Running Times in 698 Caucasian Athletes," *BMC Genomics* 19, no. 1 (2018): 1–9; Phuntila Tharabenjasin, Noel Pabalan, and Hamdi Jarjanazi, "Association of the ACTN3 R577X (rs1815739) Polymorphism with Elite Power Sports: A Meta-Analysis," *PloS one* 14, no. 5 (2019): e0217390.

21. David J. Epstein, *The Sports Gene: Inside the Science of Extraordinary Athletic Performance* (New York: Penguin, 2014), 154–157.

22. Timothy Caulfield, "Predictive or Preposterous? The Marketing of DTC Genetic Testing," *Journal of Science Communication* 10, no. 3 (2011): 3.

23. Guiseppe Lippi, Emmanuel J. Favaloro, and Gian Cesare Guidi, "The Genetic Basis of Human Athletic Performance: Why Are Psychological Components So Often Overlooked?" *Journal of Physiology* 586, no. 12 (2008): 3817.

24. Roth quoted in Larry Greenemeire, "How Olympians Could Beat the Competition by Tweaking Their Genes," *Smithsonian*, August 5, 2012, https://www.smithsonianmag.com/technology-space/how-olympians-could-beat-the-competition-by-tweaking-their-genes-14591201/#HLgF3Hj3oOXYtGYV.99. See also Craig Pickering and John Kiely, "Can Genetic Testing Predict Talent? A Case Study of 5 Elite Athletes," *International Journal of Sports Physiology and Performance* 16, no. 3 (2020): 429–434.

25. Roth quoted in Greenemeire, "How Olympians Could Beat the Competition."

26. Roger Collier, "Genetic Tests for Athletic Ability: Science or Snake Oil?" *Canadian Medical Association Journal* 184, no. 1 (2012): E43–E44.

27. Quoted in Jemma Chapman, "Gene Test for Child's Sporting Chance," *Times,* December 20, 2004.

28. Lippi, Favaloro, and Guidi, "The Genetic Basis of Human Athletic Performance," 3817.

29. Craig Pickering et al., "Can Genetic Testing Identify Talent for Sport?" *Genes* 10, no. 972 (2019), 972.

30. Quoted in Olivia Schult and Laura Rivard, "Case Study in Genetic Testing for Sports Ability," *Nature,* September 25, 2013, https://www.nature.com/scitable/forums/genetics-generation/case-study-in-genetic-testing-for-sports-107403644/.

31. Quoted in Bill Briggs, "Baby Olympian? DNA Test Screens Sports Ability," *NBC News.com,* March 4, 2009, http://www.nbcnews.com/

id/29496350/ns/health-childrens_health/t/baby-olympian-dna-test-screens-sports-ability/#.W2cTJdhKh7M.

32. Silvia Camporesi and Mike J. McNamee, "Ethics, Genetic Testing, and Athletic Talent: Children's Best Interests, and the Right to an Open (Athletic) Future," *Physiological Genomics* 48, no. 3 (2016): 191–195.

33. Julian Savulescu and Bennett Foddy, "Comment: Genetic Test Available for Sports Performance," *British Journal of Sports Medicine* 39, no. 8 (2005): 472–472.

34. Nicole Vlahovich et al., "Ethics of Genetic Testing and Research in Sport: A Position Statement from the Australian Institute of Sport," *British Journal of Sports Medicine* 51, no. 1 (2017): 5–11; Jeffrey R. Botkin et al., "Points to Consider: Ethical Legal, and Psychosocial Implications of Genetic Testing in Children and Adolescents," *American Journal of Human Genetics* 97, no. 1 (2015): 6–12.

35. Nick Webborn et al., "Direct-to-Consumer Genetic Testing for Predicting Sports Performance and Talent Identification: Consensus Statement," *British Journal of Sports Medicine* 49, no. 23 (2015): 1486; 1490.

36. World Anti-Doping Agency, *The Stockholm Declaration* (2005), cited in Andy Miah and Emma Rich, "Genetic Tests for Ability? Talent Identification and the Value of an Open Future," *Sport, Education and Society* 11, no. 3 (2006): 263.

37. Ron Synovitz and Zamira Eshanova, "Uzbekistan Is Using Genetic Testing to Find Future Olympians," *The Atlantic,* February 4, 2014, https://www.theatlantic.com/international/archive/2014/02/uzbekistan-is-using-genetic-testing-to-find-future-olympians/283001/.

38. Jason Lemon, "China Will Begin Using Genetic Testing to Select Olympic Athletes," *Newsweek,* August 31, 2018, https://www.newsweek.com/china-begin-using-genetic-testing-select-olympic-athletes-1099058.

39. Quoted in Kashmira Gander, "Nutrigenomics: Can DNA Be Used to Change Your Fitness and Diet Regimen?" *Independent,* February 16, 2017, https://www.independent.co.uk/life-style/nutrigenomics-dna-fitness-diet-regime-dnafit-does-it-work-experts-genetic-gym-health-a7582966.html.

40. Ian Varley et al., "The Association of Novel Polymorphisms with Stress Fracture Injury in Elite Athletes: Further Insights from the SFEA Cohort," *Journal of Science and Medicine in Sport* 21, no. 6 (2018): 564–568.

41. Masouda Rahim, Malcolm Collins, and Alison September, "Genes and Musculoskeletal Soft-Tissue Injuries," in *Genetics and Sports,* 2nd

ed., ed. Michael Posthumus and Malcolm Collins (New York: Karger, 2016), 68–91.

42. Shaun Assael, "Cheating Is So 1999," *ESPN,* July 10, 2012, http://www. espn.com/espn/magazine/archives/news/story?page=magazine-20101 019-article30.

43. Bobak Abdomohammadi et al., "Genetics of Chromic Traumatic Encephalopathy," *Seminars in Neurology* 40, no. 4 (2020): 426. See also Aaron J. Carman et al., "Mind the Gaps—Advancing Research into Short-Term and Long-Term Neuropsychological Outcomes of Youth Sports-Related Concussions," *Nature Reviews Neurology* 11 (2015): 230–244.

44. Quoted in Sam Peters, "England Stars Blocked RFU Concussion Gene-Testing Plan for All Professional Players Due to 'Big Brother' Privacy Fears," *Daily Mail,* November 29, 2014, https://www.dailymail.co.uk/ sport/concussion/article-2854458/England-stars-blocked-RFU-con cussion-gene-testing-plan-professional-players-Big-Brother-privacy-fears.html.

45. Stephen Smith, "What Is the Real Cost of Injuries in Professional Sport," *Kitman Labs,* April 23, 2016, https://www.kitmanlabs.com/ what-is-the-real-cost-of-injuries-in-professional-sport/.

46. Seema Patel, "Rugby, Concussions and Duty of Care: Why the Game Is Facing Scrutiny," *The Conversation,* June 3, 2021, https://theconversat ion.com/rugby-concussions-and-duty-of-care-why-the-game-is-fac ing-scrutiny-161773.

47. Andy Miah, *Genetically Modified Athletes: Biomedical Ethics, Gene Doping and Sport* (London: Routledge, 2004), 178.

48. Alexandra C. McPherron and Se-Jin Lee, "Double Muscling in Cattle Due to Mutations in the Myostatin Gene," *Proceedings of the National Academy of Sciences* 94, no. 23 (1997): 12457–12461.

49. Dana S. Mosher et al., "A Mutation in the Myostatin Gene Increases Muscle Mass and Enhances Racing Performance in Heterozygote Dogs: E79," *PLoS Genetics* 3, no. 5 (2007), e79.

50. Markus Schuelke et al., "Myostatin Mutation Associated with Gross Muscle Hypertrophy in a Child," *New England Journal of Medicine* 350, no. 26 (2004): 2682–2688. Eddie Hall, the 2017 "World's Strongest Man," claims to have the *MSTN* variant.

51. Elisabeth R. Barton-Davis et al., "Viral Mediated Expression of Insulin-like Growth Factor I Blocks the Aging-Related Loss of Skeletal Muscle Function," *Proceedings of the National Academy of Sciences* 95, no. 26 (1998): 15603–15607.

52. Melinda Wenner, "How to Be Popular during the Olympics: Be H. Lee Sweeney, Gene Doping Expert," *Scientific American* August 15, 2008, https://www.scientificamerican.com/article/olympics-gene-doping-expert/.

53. Quoted in Christen Brownlee, "Gene Doping: Will Athletes Go for the Ultimate High?" *Science News* 166, no. 18 (2004): 280.

54. Albert de la Chapelle, Ann-Liz Träskelin, and Eeva Juvonen, "Truncated Erythropoietin Receptor Causes Dominantly Inherited Benign Human Erythrocytosis," *Proceedings of the National Academy of Sciences of the United States of America* 90 (1993): 4495–4499.

55. Epstein, *The Sports Gene*, 266–281.

56. Eric C. Svensson et al., "Long-Term Erythropoietin Expression in Rodents and Non-Human Primates Following Intramuscular Injection of a Replication-Defective Adenoviral Vector," *Human Gene Therapy* 8, no. 15 (1997): 1797–1806.

57. Guangping Gao et al., "Erythropoietin Gene Therapy Leads to Autoimmune Anemia in Macaques," *Blood* 103, no. 9 (2004): 3300–3302.

58. Gretchen Reynolds, "Outlaw DNA," *New York Times,* June 3, 2007, https://www.nytimes.com/2007/06/03/sports/playmagazine/0603play-hot.html.

59. Ibid. See also Mario Thevis et al., "Trafficking of Drug Candidates Relevant for Sports Drug Testing: Detection of Non-approved Therapeutics Categorized as Anabolic and Gene Doping Agents in Products Distributed via the Internet," *Drug Testing and Analysis* 3, no. 5 (2011): 331–336.

60. Angela J. Schneider and Theodore Friedmann, *Gene Doping in Sports: The Science and Ethics of Genetically Modified Athletes* (Boston: Elsevier, 2006), xi.

61. World Anti-Doping Code, Prohibited List, 2021, accessed December 22, 2023, https://www.wada-ama.org/sites/default/files/resources/files/2021list_en.pdf.

62. Anna Baoutina, "A Brief History of the Development of a Gene Doping Test," *Bioanalysis* 12, no. 11 (2020): 723.

63. Interview with Matthew Porteus, November 8, 2021.

64. Josiah Zayner, "True Story: I Injected Myself with a CRISPR Genetic Enhancement," *The Antisense,* November 3, 2018, http://theantisense.com/2018/11/13/true-story-i-injected-myself-with-a-crispr-genetic-enhancement/; Sigal Samuel, "How Biohackers Are Trying to Upgrade Their Brains, Their Bodies—and Human Nature," *Vox,* November 15, 2019, https://www.vox.com/future-perfect/2019/6/25/18682583/bio

hacking-transhumanism-human-augmentation-genetic-engineering-crispr.

65. Porteus interview, November 8, 2021. By the 2016 Olympic Games in Rio, molecular biologist Anna Baoutina had developed a blood test to identify the use of genetically modified *EPO*. The test was put on hold because a WADA-accredited lab had not certified its use. Instead, the Olympians' blood samples were stored for later analysis.

66. David Cryanoski, "What CRISPR-Baby Prison Sentences Mean for Research," *Nature,* January 3, 2020, https://www.nature.com/articles/d41586-020-00001-y.

67. Michael Veber, "The Coercion Argument Against Performance-Enhancing Drugs," *Journal of the Philosophy of Sport* 41, no. 2 (2014): 267–277.

68. Michael Le Page, "Gene Doping in Sport Could Make the Olympics Fairer and Safer, *New Scientist,* August 5, 2016, https://www.newscientist.com/article/2100181-gene-doping-in-sport-could-make-the-olympics-fairer-and-safer/#ixzz7C2BGRZJO.

69. Miah, *Genetically Modified Athletes,* 178.

70. Verner Møller and Rasmus Bysted Møller, "Gene Doping: Ethical Perspectives," in *Routledge Handbook of Sport and Exercise System Genetics,* ed. Timothy Lightfoot, Monica J. Hubal, and Stephen M. Roth (London: Routledge, 2019), 453.

71. Julian Savulescu, "Why It's Time to Legalize Doping in Athletics," *The Conversation,* August 28, 2015, https://theconversation.com/why-its-time-to-legalise-doping-in-athletics-46514.

72. Marleen H. M. de Moor et al., "Genome-Wide Linkage Scan for Athlete Status in 700 British Female DZ Twin Pairs," *Twin Research and Human Genetics* 10, no. 6 (2007): 812–820.

73. Juliana Antero et al., "A Medal in the Olympics Runs in the Family: A Cohort Study of Performance Heritability in the Games History," *Frontiers in Physiology* 9 (2018): 1313.

74. Andrew C. Venezia and Stephen M. Roth, "The Scientific and Ethical Challenges of Using Genetics Information to Predict Sport Performance," in *Routledge Handbook of Sport and Exercise Systems Genetics,* ed. J. Timothy Lightfoot, Monica J. Hubal, and Stephen M. Roth (New York: Routledge, 2019), 443.

75. Quoted in Colin N. Moran and Guan Wang, "Genetic Limitations to Athletic Performance," in Routledge *Handbook on Biochemistry of Exercise,* ed. Peter M. Tiidus et al. (London: Routledge, 2020), 217–231.

76. "Are You Elite?," ELITE, Sanford Online, accessed December 22, 2023, https://elite.stanford.edu/.

77. Quoted in Jon Wilner, "Can Superhuman Athletes Provide Genetic Clues on Heart Health?" *Mercury News,* October 29, 2017, https://www.mercurynews.com/2017/10/29/4851089/.

CONCLUSION

1. "WADA Statement on Russian Anti-Doping Agency Finding of 'No Fault or Negligence' in Case of ROC Figure Skater," January 13, 2023, https://www.wada-ama.org/en/news/wada-statement-russian-anti-doping-agency-finding-no-fault-or-negligence-case-roc-figure.

2. "WADA Appeals Case of Russian Olympic Committee Figure Skater to Court of Arbitration for Sport," February 21, 2023, https://www.wada-ama.org/en/news/wada-appeals-case-russian-olympic-committee-figure-skater-court-arbitration-sport.

3. International Skating Union, "#FigureSkating," February 22, 2023, https://www.isu.org/isu-news/news/145-news/14500-isu-appeals-case-of-roc-figure-skater-to-court-of-arbitration-for-sport?templatePa ram=15.

4. Court of Arbitration for Sport, "Media Release," January 29, 2024, https://www.tascas.org/fileadmin/user_upload/CAS_Media_Rele ase_9451_9455_9456_Decision.pdf

5. Ibid.

6. Helen Jefferson Lenskyj, *The Olympic Games: A Critical Approach* (Leeds: Emerald Insight, 2020), 108, and *Gender, Athletes' Rights, and the Court of Arbitration for Sport* (Leeds: Emerald Insight, 2018).

7. Court of Arbitration for Sport, CAS 2020/A/6807 Blake Leeper v. International Association of Athletics Federations, accessed December 21, 2023, https://www.tas-cas.org/fileadmin/user_upload/Award_ _6807___for_publication_.pdf.

8. Seema Patel, "Gaps in the Protection of Athletes Gender Rights in Sport—A Regulatory Riddle," *International Sports Law Journal* 21 (2021): 265.

9. Bruce Kidd and Peter Donnelly, "Human Rights in Sports," *International Review for the Sociology of Sport* 35, no. 2 (2000): 135.

10. Daniela Heerdt, "The Court of Arbitration for Sport: Where Do Human Rights Stand?" Centre for Sport and Human Rights, May 10, 2019, https://sporthumanrights.org/library/the-court-of-arbitration-for-sport-where-do-human-rights-stand/ .

11. Lena Holzer, "What Does It Mean to Be a Woman in Sports? An Analysis of the Jurisprudence of the Court of Arbitration for Sport," *Human Rights Law Review* 20, no. 3 (2020): 405.

12. United Nations, Convention on the Rights of the Child (1989), https://www.ohchr.org/en/instruments-mechanisms/instruments/convention-rights-child. See Paulo David, *Human Rights in Youth Sport: A Critical Review of Children's Rights in Competitive Sport* (London: Routledge, 2004); Siri Farstad, "Protecting Children's Rights in Sport: The Use of Minimum Age," *Human Rights Law Commentary* 3 (2007): 1–20.

13. John G. Ruggie, *"For the Game. For the World." FIFA and Human Rights*, Corporate Responsibility Initiative Report No. 68 (Cambridge, MA: Harvard Kennedy School, 2016), 36.

14. WADA, "The World Anti-Doping Code," accessed December 21, 2023, https://www.wada-ama.org/en/what-we-do/world-anti-doping-code.

15. International Paralympic Committee, accessed December 21, 2023, "Classification Code," https://www.paralympic.org/classification-code.

16. Frank de Zwart, "Unintended but Not Unanticipated Consequences," *Theoretical Sociology* 44 (2015): 286–287.

17. International Skating Union, *Agenda of the 58th Ordinary Congress, Phuket*, 2022, https://www.isu.org/docman-documents-links/isu-files/documents-communications/about-isu/congress-documents/28312-isu-communication-2472-1/file.

18. Christine Brennan, *Inside Edge: A Revealing Journey into the Secret World of Figure Skating* (New York: Scribner, 1996), 31.

19. Sandra Loosemore, "'Figures' Don't Add Up in Competition Anymore," *CBS SportsLine,* December 16, 1998, https://web.archive.org/web/20080727021537/http://cbs.sportsline.com/u/women/skating/dec98/loosemore121698.htm.

20. "No More Figures in Figure Skating," *New York Times*, June 9, 1988, https://www.nytimes.com/1988/06/09/sports/no-more-figures-in-figure-skating.html#:~:text=The%20International%20Skating%20Union%20voted,be%20reduced%20in%20dance%20programs; James R. Hines, *Figure Skating: A History* (Urbana: University of Illinois Press, 2006), 205. Jumps became even more important when the ISU overhauled its scoring system in 2004.

21. Lenskyj, *Gender, Athletes' Rights*, 162.

22. Stéphane Bermon et al., "Serum Androgen Levels Are Positively Correlated with Athletic Performance and Competition Results in Elite Female Athletes," *BJSM* 52, no. 23 (2018): 1531–1532.

23. Jonathan Ruwuya, Byron Omwando Juma, and Jules Woolf, "Challenges Associated with Implementing Anti-Doping Policy and Programs in Africa," *Frontiers in Sports and Active Living* 4 (December 8, 2022): 966559, doi:10.3389/fspor.2022.966559

24. Roger Pielke, Jr., *The Edge: The War Against Cheating and Corruption in the Cutthroat World of Elite Sport* (Berkeley, CA: Roaring Forties Press, 2016), 269.

25. Angela Schneider, "Banned from the Tokyo Olympics for Pot? Let the Athletes Decide What Drugs Should Be Allowed," *The Conversation*, July 6, 2021, https://theconversation.com/banned-from-the-tokyo-olympics-for-pot-let-the-athletes-decide-what-drugs-should-be-allowed-163619. See also Emmanuel Macedo et al., "Moral Communities in Anti-Doping Policy: A Response to Bowers and Paternoster," *Sport, Ethics and Philosophy* 13, no. 1 (2019): 49–61.

26. Lauryn Stewart et al., "Developing Trans-athlete Policy in Australian National Sport Organizations," *International Journal of Sport Policy and Politics* 13, no. 4 (2021): 565–585.

27. Phillip L. Craven, "The Development from a Medical Classification to a Player Classification in Wheelchair Basketball," *Adapted Physical Activity* (1990): 81.

28. While several nations and sports governing bodies have established these kinds of reporting systems, they have limited reach and are sometimes flawed. See, for example, Joseph John, Gretchen Kerr, and Simon Darnell, "'Safe Sport Is Not for Everyone': Equity-Deserving Athletes' Perspectives of, Experiences, and Recommendations for Safe Sport," *Frontiers in Psychology* (2022), doi.org/10.3389/fpsyg.2022.832560.

29. Yetsa A. Tuakli-Wosornu et al., "The Journey to Reporting Child Protection Violations in Sport: Stakeholder Perspectives," *Frontiers in Psychology* 13 (2022), https://doi.org/10.3389/fpsyg.2022.907247 https://www.frontiersin.org/articles/10.3389/fpsyg.2022.907247/full

30. "WADA Launches 'Speak Up!'—A Secure Digital Platform to Report Doping Violations," March 9, 2017, https://www.wada-ama.org/en/news/wada-launches-speak-secure-digital-platform-report-doping-violations.

31. World Anti-Doping Agency, "Frequently Asked Questions," accessed December 21, 2023, https://www.wada-ama.org/sites/default/files/resources/files/athlete_central_faq_final_en_0.pdf.

32. Quoted in Talya Minsberg, "In Her New Book, Kara Goucher Keeps Running Accountable," *New York Times*, March 25, 2023, https://www.nytimes.com/2023/03/25/sports/running-goucher-longest-race.html.

33. Joan Ryan, "When Will Child Athletes Be Protected?" *Washington Post,* February 21, 2022, A17.

34. Franklin Foer, "The Goals of Globalization," *Foreign Policy* 12, no. 5 (2005), https://foreignpolicy.com/2009/10/20/the-goals-of-global ization/

35. David S. Heineman, "Leaving Fandom: Why I Gave Up Sports, Why You Should Consider It, and How to Start," *Medium,* September 21, 2016, https://medium.com/@DrHeineman/leaving-fandom-why-i-gave-up-sports-why-you-should-consider-it-and-how-to-start-6d194 1c1a915.

36. Gina Wright, the host of the "She Talks Football" channel on YouTube.

Bibliography

"20-Year Old Para-Swimmer Ailbhe Kelly Announces Retirement." *SwimSwam,* September 15, 2019. https://swimswam.com/20-year-old-para-swimmer-ailbhe-kelly-announces-retirement/.

Abdomohammadi, Bobak, Alicia Dupre, Laney Evers, and Jesse Mez. "Genetics of Chronic Traumatic Encephalopathy." *Seminars in Neurology* 40, no. 4 (2020): 426.

Abkowitz, Janis L. "President's Column—Sickle Cell Trait and Sports: Is the NCAA a Hematologist?" *The Hematologist* 10, no. 3 (2013). https://doi.org/10.1182/hem.V10.3.1070.

Akkermans, Rebecca. "Ludwig Guttmann." *The Lancet Neurology* 15, no. 12 (2016): 1210.

Allan, Veronica. "Do Para-Athletes Face Abuse Trying to Prove Their Disabilities." *News Decoder,* March 13, 2019. https://news-decoder.com/2019/03/13/para-athletes-classification-abuse/.

American College of Sports Medicine. "Position Stand on Weight Loss in Wrestlers." *Medicine and Science in Sports and Exercise* 8, no. 2 (1976): xi–xiii.

Amy-Chinn, Dee. "The Taxonomy and Ontology of Sexual Difference: Implications for Sport." *Sport in Society* 15, no. 9 (2012): 1291–1305.

Anderson, Eric, and Jennifer Hargreaves, eds. *Routledge Handbook of Sport, Gender and Sexuality.* London: Routledge, 2014.

Anderson, Eric, and Ann Travers, eds. *Transgender Athletes in Competitive Sport.* London: Taylor & Francis, 2017.

Anderson, Jack. "A Brief Legal History of Prize Fighting in Nineteenth Century America." *Sport in History* 24, no. 1 (2004): 32–62.

Antero, Juliana, Guillaume Saulière, Adrien Marck, and Jean-François Toussaint. "A Medal in the Olympics Runs in the Family: A Cohort Study of Performance Heritability in the Games History." *Frontiers in Physiology* 9 (2018): 1313.

"Anti-Doping Agency Won't Ban Oxygen Tents." *CBC Sports,* September 16, 2006. https://www.cbc.ca/sports/anti-doping-agency-won-t-ban-oxygen-tents-1.589665.

Archibald, Catherine Jean. "Transgender and Intersex Sports Rights." *Virginia Journal of Social Policy & Law* 26, no. 3 (2019): 246–276.

Ardigò, Luca Paolo, Wissem Dhabhi, and Johnny Padulo. "Height-Based Model for the Categorization of Athletes in Combat Sports." *European Journal of Sport Science* (2020): 1–10.

Armstrong, David J. "Hypoxic Chambers and Other Artificial Environments." In *Drugs in Sport*, edited by David R. Mottram, 354–369. London: Routledge, 2012.

Artioli, Guilherme Giannini, Bruno Gualano, Emerson Franchini, Fernanda Baeza Scagliusi, Mariane Takesian, Marina Fuchs, and Antonio Herbert Lancha. "Prevalence, Magnitude, and Methods of Rapid Weight Loss Among Judo Competitors." *Medicine and Science in Sports and Exercise* 42 (2010): 436–442.

Artioli, Guilherme Giannini, Bryan Saunders, Rodrigo T. Iglesias, and Emerson Franchini. "Authors' Reply to Davis: "It Is Time to Ban Rapid Weight Loss from Combat Sports." *Sports Medicine* 47, no. 8 (2017): 1677–1681.

Assael, Shaun. "Cheating is so 1999." *ESPN,* July 10, 2012. http://www.espn. com/espn/magazine/archives/news/story?page=magazine-20101019-article30.

Auchus, Richard J. "Endocrinology and Women's Sports: The Diagnosis Matters." *Law and Contemporary Problems* 80, no. 127 (2018): 135–136.

Bachynski, Kathleen. *No Game for Boys to Play: The History of Youth Football and the Origins of a Public Health Crisis.* Chapel Hill: University of North Carolina Press, 2019.

Bailey, Steve. *Athlete First: A History of the Paralympic Movement.* Hoboken, NJ: Wiley, 2008.

Bairner, Alan. "An Introduction to the Negative Aspects of Sport." *Idrotts Forum,* August 29, 2022. https://idrottsforum.org/baiala_anderson-magrath220829/.

Balmer, Nigel, Pascoe Pleasence, and Alan Nevill. "Evolution and Revolution: Gauging the Impact of Technological and Technical Innovation on Olympic Performance." *Journal of Sports Science* 30 (2011): 1075–1083.

Bantjes, Jason, and Leslie Swartz. "Social Inclusion through Para Sport: A Critical Reflection on the Current State." *Physical Medicine and Rehabilitation Clinics of North America* 29 (2018): 409–416.

Baoutina, Anna. "A Brief History of the Development of a Gene Doping Test." *Bioanalysis* 12, no. 11 (2020): 723–727.

Barley, Oliver R., Dale W. Chapman, and Chris R. Abbiss. "Weight Loss Strategies in Combat Sports and Concerning Habits in Mixed Martial Arts." *International Journal of Sports Physiology and Performance* (2017): 1–24.

Barnes, Dan. "'I'm Number One in the World—Other Than Ukraine': Eastern Euro Trio Thwarts Legendary Canadian Swimmer." *National Post,* September 12, 2016. https://nationalpost.com/sports/olympics/im-num ber-one-in-the-world-other-than-ukraine-eastern-european-trio-thwa rts-legendary-canadian-swimmer.

Barnes, Katie. "IOC Provides Framework for International Federations to Develop Their Own Eligibility Criteria for Transgender, Intersex Athletes." *ESPN.com,* November 16, 2021. https://www.espn.com/olymp ics/story/_/id/32645620/ioc-provides-framework-international-federati ons-develop-their-own-eligibility-criteria-transgender-intersex-athletes.

Baron, David A., David M. Martin, and Samir Abol Magd. "Doping in Sports and Its Spread to At-Risk Populations: An International Review." *World Psychiatry* (2007): 118–123.

Barton-Davis, Elisabeth R., Daria I. Shoturma, Antonio Musaro, Nadia Rosenthal, and H. Lee Sweeney. "Viral Mediated Expression of Insulin-like Growth Factor I Blocks the Aging-related Loss of Skeletal Muscle Function." *Proceedings of the National Academy of Sciences* 95, no. 26 (1998): 15603–15607.

Batelaan, Krystal, and Gamal Abdel-Shehid. "On the Eurocentric Nature of Sex Testing: The Case of Caster Semenya." *Social Identities* 27, no. 2 (2021): 146–165.

Beckman, Emma M., Mark J. Connick, Mike J. McNamee, Richard Parnell, and Sean M. Tweedy. "Should Markus Rehm Be Permitted to Compete in the Long Jump at the Olympic Games?" *British Journal of Sports Medicine* 51, no. 14 (2017): 1048–1049.

Bederman, Gail. *Manliness and Civilization: A Cultural History of Gender and Race in the United States, 1880–1917.* Chicago: University of Chicago Press, 2008.

Berlioux, Monique. "Femininity." *Olympic Review* 3 (December 1967): 1–2.

Bermon, Stéphane, Pierre Yves Garnier, Angelica Lindén Hirschberg, Neil Robinson, Sylvain Giraud, Raul Nicoli, Norbert Baume, Martial Saugy, Patrick Fénichel, Stepen J. Bruce Hugues Henry, Gariel Dollé, and Martin Ritzen. "Serum Androgen Levels in Elite Female Athletes." *Journal of Clinical Endocrinology and Metabolism* 99, no. 11 (2014): 4328–4335.

Bermon, Stéphane, Angelica Lindén Hirschberg, Jan Kowalski, and Emma Eklund. "Serum Androgen Levels Are Positively Correlated with Athletic

Performance and Competition Results in Elite Female Athletes." *BJSM* 52, no. 23 (2018): 1531–1532.

Bermon, Stéphane, Martin Ritzen, Angelica Lindén Hirschberg, and Thomas H. Murray. "Are the New Policies on Hyperandrogenism in Elite Female Athletes Really Out of Bounds?" *American Journal of Bioethics* 13, no. 5 (2013): 63–65.

Berryman, Jack W. *Out of Many, One: A History of the American College of Sports Medicine.* Champaign, IL: Human Kinetics, 1995.

Berthelot, Geoffroy, Adrien Sedeaud, Adrien Marck, Juliana Antero-Jacquemin, Julien Schipman, Guillaume Sauliere, Andy Marc, François-Denis Desgorces, and Jean-François Toussaint. "Has Athletic Performance Reached its Peak?" *Sports Medicine* 45, no. 9 (2015): 1263–1271.

Berthelot, Geoffroy, Muriel Tafflet, Nour El Helou, Stephane Len, Sylvie Escolano, Marion Guillaume, Hala Nassif, Julien Tolaïni, Valérie Thibault, François-Denis Desgorces, Olivier Hermine, and Jean-Francois Toussaint. "Athlete Atypicity on the Edge of Human Achievement: Performances Stagnate after the Last Peak in 1988." *PLoS ONE* 5 (2010): e8800 doi: 10.1371/journal.pone.0008800.

Berthelot, Geoffroy, Valérie Thibault, Muriel Tafflet, Sylvie Escolano, Nour El Helou, Xavier Jouven, Olivier Hermine, and Jean-François Toussaint. "The Citius End: World Records Progression Announces the Completion of a Brief Ultra-Physiological Quest." *PLoS ONE* 3 (2008): e1552 doi: 10.1371/journal.pone.0001552.

Best, Krista L., W. Ben Mortenson, Zach Lauzière-Fitzgerald, and Emma M. Smith. "Language Matters! The Long-standing Debate between Identity-first Language and Person First Language." *Assistive Technology* 34, no. 2 (2022): 127–128.

Betancurt, Jonathan Ospina, Maria S. Zakynthinaki, Maria Jose Martinez-Patiño, and Carlos Cordente Martinez. "Hyperandrogenic Athletes: Performance Differences in Elite-Standard 200m and 800m Finals." *Journal of Sports Sciences* 36, no. 21 (2018): 2464–2471.

Bezuglov, Eduard, Oleg Talibov, Mikhail Butovskiy, Vladimir Khaitin, Evgeny Achkasov, Zbigniew Waśkiewicz, and Artemii Lazarev. "The Inclusion in WADA Prohibited List is Not Always Supported by Scientific Evidence: A Narrative Review." *Asian Journal of Sports Medicine* 12, no. 2 (2021): e110753.

Bhowmick, Nilanjana and Jyoti Thottam. "Gender and Athletics: India's Own Caster Semenya." *Time*, September 1, 2009. http://content.time.com/time/world/article/0,8599,1919562,00.html.

Bianchi, Andria. "Something's Got to Give: Reconsidering the Justification for a Gender Divide in Sport." *Philosophies* 4, no. 2 (2019): https://doi.org/10.3390/philosophies4020023.

Bock, William. "The Effects of Dehydration upon the Cardio-Respiratory Endurance of Wrestlers." PhD diss., Ohio State University, 1965.

Boddy, Kasia. "'Under the Queensberry Rules, So to Speak': Some Versions of a Metaphor." *Sport in History*, 31, no. 4 (2011): 398–422.

Bohn, Mike. "UFC's Cris 'Cyborg': Why She's Forced to Lose Weight to Win." *Rolling Stone,* September 22, 2016. https://www.rollingstone.com/culture/culture-sports/ufcs-cris-cyborg-why-shes-forced-to-lose-weight-to-win-187532/.

Boin, Arjen, Lauren A. Fahy, Paul 't Hart, eds. *Guardians of Public Value: How Public Organizations Become and Remain Institutions.* Cham, Switzerland: Palgrave Macmillan, 2021.

Bøje, Ove. "Doping: A Study of the Means Employed to Raise the Level of Performance in Sport." *Bulletin of the Health Organization of the League of Nations* 8 (1939): 439.

Botkin, Jeffrey R., John W. Belmont, Jonathan S. Berg, Benjamin E. Berkman, Yvonne Bombard, Ingrid A. Holm, Howard P. Levy, Kelly E. Ormond, Howard M. Saal, Nancy B. Spinner, Benjamin S. Wilfond, and Joseph D. McInerny. "Points to Consider: Ethical Legal, and Psychosocial Implications of Genetic Testing in Children and Adolescents." *American Journal of Human Genetics* 97, no. 1 (2015): 6–12.

Bower, Moisekapenda. "Parents of Former Rice DB File Wrongful Death Suit." *Chronicle,* August 9, 2011. https://www.chron.com/sports/college-football/article/Parents-of-former-Rice-DB-file-wrongful-death-suit-1788862.php.

Bowman, Verity, and Luke Mintz. "The Dark Truth behind the Beauty of Figure Skating." *Daily Telegraph,* February 11, 2022.

Bray, Molly S., James M. Hagberg, Louis Perusse, Tuomo Rankinen, Stephen M. Roth, Bernd Wolfarth, and Claude Bouchard. "The Human Gene Map Performance and Health-Related Fitness Phenotypes: The 2006–2007 Update." *Medicine and Science in Sports and Exercise* 41, no. 1 (2009): 35–73.

Brennan, Christine. *Inside Edge: A Revealing Journey into the Secret World of Figure Skating.* New York: Scribner, 1996.

Briggs, Bill. "Baby Olympian? DNA Test Screens Sports Ability." *NBC News. com,* March 4, 2009. https://www.nbcnews.com/health/health-news/baby-olympian-dna-test-screens-sports-ability-flna1c9451428.

Brissonneau, Christophe. "The 1998 Tour de France: Festina, from Scandal to an Affair in Cycling." In *Routledge Handbook of Drugs and Sport,* edited by Verner Møller, Ivan Waddington, and John M. Hoberman, 181–192. London: Routledge, 2015.

Brittain, Ian. *The Paralympic Games Explained*. London: Routledge, 2016.

Brittain, Ian, and Aaron Beacom, eds. *The Palgrave Handbook of Paralympic Studies*. London: Palgrave Macmillan, 2018.

Brito, Ciro José, Aendria Fernanada Castro Martins Roas, Igor Surian Souza Brito, João Carlows Bouzas Marins, Claudio Córdova, and Emerson Franchini. "Methods of Body Mass Reduction by Combat Sport Athletes." *International Journal of Sport Nutrition and Exercise Metabolism* 22 (2012): 89–97.

Brown, Andy. "DSD & Transgender Athletes: Paula Radcliffe's View." *Sports Integrity Initiative,* April 30, 2019. https://www.sportsintegrityinitiative.com/dsd-transgender-athletes-paula-radcliffes-view/.

Brown, Andy. "Paralympic Classification System Allegedly Open to Abuse." *Sport Integrity Initiative,* November 2, 2017. https://www.sportsintegrityinitiative.com/paralympic-classification-system-allegedly-open-abuse/.

Browne, Alister, Victor Lachance, and Andrew Pipe. "The Ethics of Blood Testing as an Element of Doping Control in Sport." *Medicine and Science in Sports and Exercise* 31, no. 4 (1999): 497–501.

Brownlee, Christen. "Gene Doping: Will Athletes Go for the Ultimate High?" *Science News* 166, no. 18 (2004): 280.

Buchanan, Benjamin K., David M. Siebert, Monica L. Ziamgan Suchsland, Jonathan A. Drezner, Irfan M. Asif, Francis G. O'Connor, and Kimberly G. Harmon. "Sudden Deaths Associated with Sickle Cell Trait Before and After Mandatory Screening." *Sports Health: A Multidisciplinary Approach* 12, no. 3 (2020). https://doi.org/10.1177/1941738120915690.

Burke, Louis M., Gary J. Slater, Joseph J. Matthews, Carl Langan-Evans, and Craig A. Horswill. "ACSM Expert Consensus Statement of Weight Loss in Weight-Category Sports." *Current Sports Medicine Reports* 20, no. 4 (2021): 199–271.

Burkett, Brendan, Carl Payton, Peter Van de Vliet, Hannah Jarvis, Daniel Daly, Christiane Mehrkuehler, Marvin Kilian, and Luke Hogarth. "Performance Characteristics of Para Swimmers: How Effective is the Swimming Classification System?" *Physical Medicine and Rehabilitation Clinics* 29, no. 2 (2018): 333–346.

Buts, Caroline, Cind Du Boise, Bruno Heyndels, and Marc Jegers. "Socioeconomic Determinants of Success at the Summer Paralympics." *Journal of Sports Economics* 14, no. 2 (2011): 133–147.

Cahn, Susan K. *Coming on Strong: Gender and Sexuality in Women's Sport*. Urbana: University of Illinois Press, 2015.

Camporesi, Silvia, and Mike J. McNamee. "Ethics, Genetic Testing, and Athletic Talent: Children's Best Interests, and the Right to an Open (Athletic) Future." *Physiological Genomics* 48, no. 3 (2016): 191–195.

Caplan-Bricer, Nora. "The Inextricable Tie Between Eating Disorders and Endurance Athletes." *Outside*, June 23, 2017. https://www.outsideonline. com/2191906/eating-disorders-are-more-common-you-think#:~:text= %E2%80%9CBeing%20a%20professional%20athlete%20isn,much%20 a%20body%20can%20withstand.%E2%80%9D.

Carl, Rebecca L., Miriam D. Johnson, and Thomas J. Martin. "Promotion of Healthy Weight-Control Practices in Young Athletes." *Pediatrics* 140, no. 3 (2017): e20171871. https://doi.org/10.1542/peds.2017-1871.

Carlson, Alison. "When Is a Woman Not a Woman?" *Women's Sport & Fitness*, March 1991, 29.

Carman, Aaron J., Rennie Ferguson, Robert Cantu, R. Dawn Comstock, Penny A. Dacks, Steven T. DeKosky, Sam Gandy, James Gilbert, Chad Gilliand, Gerard Gioia, Christopher Giza, Michael Grecius, Brian Hainline, Ronald L. Hayes, James Hendrix, Barry Jordan, James Kovach, Rachel F. Lane, Rebekah Mannix, Thomas Murray, Tad Seifert, Diana W. Shineman, Eric Warren, Elisabeth Wilde, Huntington Willard, and Howard M. Fillit. "Mind the Gaps—Advancing Research into Short-Term and Long-Term Neuropsychological Outcomes of Youth Sports-Related Concussions." *Nature Reviews Neurology* 11 (2015): 230–244.

Carter, Neil. "Monkey Glands and the Major: Frank Buckley and Modern Football Management." Manchester Metropolitan University, 2011. https://dora.dmu.ac.uk/bitstream/handle/2086/4617/Frank%20Buck ley%20Essay.pdf?sequence=3.

Cashmore, Ellis. "Opinion: It's Time to Allow Doping in Sport." *CNN*, October 24, 2012. https://www.cnn.com/2012/10/23/opinion/cashm ore-time-to-allow-doping-in-sport/index.html.

Castilha, Fabio Andre, Heros Ribeiro Ferreira, Glauber Oliveira, Talita Oliveira, Paula Roquetti Fernandes, and Jose Fernandes Filho. "The Influence of Gene Polymorphisms and Genetic Markers in the Modulation of Sports Performance: A Review." *Journal of Exercise Physiology Online* 21, no. 2 (2018): 248–265.

Catlin, Don H., Kenneth D. Fitch, and Arne Ljungqvist. "Medicine and Science in the Fight Against Doping in Sport." *Journal of Internal Medicine* 264, no. 2 (2008): 99–114.

Caulfield, Timothy. "Predictive or Preposterous? The Marketing of DTC Genetic Testing." *Journal of Science Communication* 10, no. 3 (2011): 1–6.

Cavanaugh, Sheila L. and Heather Sykes. "Transsexual Bodies at the Olympics: The International Olympic Committee's Policy on Transsexual Athletes at the 2004 Athens Summer Games." *Body & Society* 12, no. 3 (2006): 75–102.

Chapman, Jemma. "Gene Test for Child's Sporting Chance." *Times, December* 20, 2004.

Chapman, Mike. *Encyclopedia of American Wrestling.* Champaign, IL: Human Kinetics, 1990.

Chavkin, Daniel. "Russia Responds to Criticism from IOC President About Valieva's Coach." *Sports Illustrated,* February 18, 2022. https://www.si.com/olympics/2022/02/18/kamila-valieva-coach-kremlins-spokes man-olympics-president.

Chiari, Mike. "Kevin Lee Says He Didn't Know Where He Was during UFC 216 Weight Cut." *Bleacher Report,* October 12, 2017. https://bleacherrep ort.com/articles/2738367-kevin-lee-says-he-didnt-know-where-he-was-during-ufc-216-weight-cut.

Christesen, Paul, and Donald G. Kyle, eds. *Companion to Sport and Spectacle in Greek and Roman Antiquity.* Chichester: Wiley Blackwell, 2013.

"Chromosomes Do Not an Athlete Make." *Journal of the American Medical Association* 202, no. 11 (1967): 54–55.

"CIRC Report: Executive Summary." March 2015. https://www.velonews.com/2015/03/news/circ-report-executive-summary_362351.

Clark, Richard V., Jeffrey A. Wald, Ronald S. Swerdloff, Christina Wang, Frederick CW Wu, Larry D. Bowers, and Alvin M. Matsumoto. "Large Divergence in Testosterone Concentrations between Men and Women: Frame of Reference for Elite Athletes in Sex-specific Competition in Sports, A Narrative Review." *Clinical Endocrinology* 90, no. 1 (2019): 15–22.

Coleman, Doriane Lambelet. "Sex, Sport, and Why Track and Field's New Rules on Intersex Athletes Are Essential." *New York Times,* April 30, 2018. https://www.nytimes.com/2018/04/30/sports/track-gender-rules.html.

Coleman, Doriane Lambelet, James E Coleman, Paul H. Haagen, and Curtis A. Bradley. "Position Paper of the Center for Sports Law and Policy: Whether Artificially Induced Hypoxic Conditions Violate the 'Spirit of Sport.'" Center for Sports Law and Policy, Duke Law, 2006. https://law.duke.edu/features/pdf/hypoxiaresponse.pdf.

Collier, Roger. "Genetic Tests for Athletic Ability: Science or Snake Oil?" *Canadian Medical Association Journal* 184, no. 1 (2012): E43–E44.

Collins, Tony. *Sport in a Capitalist Society: A Short History.* London: Routledge, 2013.

Connick, Mark J., Emma Beckman, and Sean M. Tweedy. "Evolution and Development of Best Practice in Paralympic Classification." In *The Palgrave Handbook of Paralympic Studies*, edited by Ian Brittain and Aaron Beacom, 389–416. London: Palgrave Macmillan, 2018.

Connor, John, and Brendan Egan. "Comparison of Hot Water Immersion at Self-Adjusted Maximum Tolerance Temperature, with or without the Addition of Salt, for Rapid Weight Loss in Mixed Martial Arts." *Biology of Sport* 38, no. 1 (2021): 89–96.

Cooper, Chris E. *Run, Swim, Throw, Cheat: The Science Behind Drugs in Sport.* New York: Oxford University Press, 2012.

"Correction: Serum Androgen Levels and Their Relation to Performance in Track and Field: Mass Spectrometry Results from 2127 Observations in Male and Female Elite Athletes." *British Journal of Sports Medicine* 55 (2021): e7.

Court of Arbitration for Sport. *Arbitral Award Delivered by the Court of Arbitration for Sport, CAS 2018/O/5794, Mokgadi Caster Semenya v. International Association of Athletics Federations; CAS 2018/O/5798 Athletics South Africa v. International Association of Athletics Federations.* 2019. https://www.tas-cas.org/fileadmin/user_upload/CAS_Award_-_redacted_-_Semenya_ASA_IAAF.pdf.

Court of Arbitration for Sport. *Arbitration CAS 2008/A/1480 Pistorius v/ IAAF, award of 16 May 2008.* Accessed December 22, 2023. https://jurisprudence.tas-cas.org/Shared%20Documents/1480.pdf.

Court of Arbitration for Sport. CAS 2020/A/6807 Blake Leeper v. International Association of Athletics Federations. 2020. https://www.tas-cas.org/fileadmin/user_upload/Award__6807___for_publication_.pdf.

Court of Arbitration for Sport. *Interim Arbitral Award: Dutee Chand v. Athletics Federation of India and the International Association of Athletics Federation, CAS 2014/A/3759.* July 24, 2015. https://jurisprudence.tas-cas.org/Shared%20Documents/3759-PA.pdf.

Court of Arbitration for Sport. "Media Release: The Court of Arbitration for Sport (CAS) Partially Upholds the Appeal of Blake Lepper." October 26, 2020. https://www.tas-cas.org/fileadmin/user_upload/CAS_Media_Release_6807.pdf.

Court of Arbitration for Sport. "Semenya, ASA and IAAF: Executive Summary." May 1, 2019. https://www.tas-cas.org/fileadmin/user_upload/CAS_Executive_Summary__5794_.pdf.

"Crash Diets for Athletes Termed Dangerous." *Journal of the American Medical Association Health Bulletin* (February 1959): 8–9.

Craven, Phillip L. "The Development from a Medical Classification to a Player Classification in Wheelchair Basketball." In *Adapted Physical Activity: An Interdisciplinary Approach*, edited by Kyonosuke Yabe, 81–86. Heidelberg: Springer Berline, 1990.

Crighton, Ben, Graeme L. Close, and James P. Morton. "Alarming Weight Cutting Behaviours in Mixed Martial Arts: A Cause for Concern and a Call for Action." *British Journal of Sports Medicine* 50, no. 8 (2016): 446–447.

"CRISPR-Cas9 Gene Editing for Sickle Cell Disease and β-Thalassemia." *New England Journal of Medicine* 384 (2021): 252–260.

Cryanoski, David. "What CRISPR-Baby Prison Sentences Mean for Research." *Nature,* January 3, 2020. https://www.nature.com/articles/d41 586-020-00001-y.

Czerniawski, Amanda M. "From Average to Ideal: The Evolution of the Height and Weight Table in the United States, 1836–1943." *Social Science History* 31, no. 2 (2007): 273–296.

Darcy, Simon. "The Paralympic Movement: A Small Number of Behemoths Overwhelming a Large Number of Also-Rans—A Pyramid Built on Sand." In *The Palgrave Handbook of Paralympic Studies*, edited by Ian Brittain and Aaron Beacom, 221–246. London: Palgrave Macmillan, 2018.

David, Paulo. *Human Rights in Youth Sport: A Critical Review of Children's Rights in Competitive Sport*. London: Routledge, 2004.

Davis, Lennard, and David Morris, "The Biocultures Manifesto." In *The End of Normal: Identity in A Biocultural Era*, edited by Lennard Davis, 121–128. Ann Arbor: University of Michigan Press, 2014.

Davison, C. St. C. B. "Landmarks in the History of Weighing and Measuring." *Transactions of the Newcomen Society* 31, no. 1 (1957): 131–152.

De Hon, Olivier, Harm Kuipers, and Maarten van Bottenburg. "Prevalence of Doping Use in Elite Sports: A Review of Numbers and Methods." *Sports Medicine,* 45, no. 1 (2015): 57–69.

de la Chapelle, Albert, Ann-Liz Träskelin, and Eeva Juvonen. "Truncated Erythropoietin Receptor Causes Dominantly Inherited Benign Human Erythrocytosis." *Proceedings of the National Academy of Sciences of the United States of America* 90 (1993): 4495–4499.

De Moor, Marleen H. M., Tim D. Spector, Lynn F. Cherkas, Mario Falchi, Jouke Jan Hottenga, Dorret I. Boomsma, and Eco J. C. De Geus. "Genome-Wide Linkage Scan for Athlete Status in 700 British Female DZ Twin Pairs." *Twin Research and Human Genetics* 10, no. 6 (2007): 812–820.

de Zwart, Frank. "Unintended but Not Unanticipated Consequences." *Theoretical Sociology* 44 (2015): 283–297.

DeBenedette, Valerie. "For Jockeys, Injuries Are Not a Long Shot." *The Physician and Sportsmedicine* 15, no. 6 (1987): 236–245.

Demos, Vasilikie, and Marcia Texler Segal, eds. *Gender Panic, Gender Policy*. Leeds: Emerald Insight, 2017.

Dennis, Mike. "The East German Doping Programme." In *Routledge Handbook of Drugs and Sport*, edited by Verner Møller, Ivan Waddington, and John Hoberman, 170–180. London: Routledge, 2015.

Dennis, Mike. "Securing the Sports 'Miracle': The Stasi and East German Elite Sport." *International Journal of the History of Sport* 29, no. 18 (2012): 2551–2574.

Denny, Mark W. "Limits to Running Speed in Dogs, Horses and Humans." *Journal of Experimental Biology* 211 (2008), 3836–3849.

DePauw, Karen P. "A Historical Perspective of the Paralympic Games." *Journal of Physical Education, Recreation & Dance* 83, no. 3 (2012): 21–31.

DePauw, Karen P., and Susan J. Gavron. *Disability Sport*. 2nd ed. Champaign, IL: Human Kinetics, 2005.

Dhanaphatana, Sarah. "Freshman Swimmer Beats the Odds." *Daily Trojan* (University of Southern California), September 28, 2014. https://dailytro jan.com/2014/09/28/freshman-swimmer-beats-the-odds/.

Dimeo, Paul. *A History of Drug Use in Sport 1876–1976: Beyond Good and Evil*. London: Routledge, 2008.

Dimeo, Paul, Thomas M. Hunt, and Richard Horbury. "The Individual and the State: A Social Historical Analysis of the East German 'Doping System.'" *Sport in History* 31, no. 2 (2011): 218–237.

Dimeo, Paul, and Verner Møller. *The Anti-Doping Crisis in Sport: Causes, Consequences, Solutions*. London: Routledge, 2018.

Dingeon, Bernard. "Gender Verification and the Next Olympic Games." *Journal of the American Medical Association*, 269, no. 3 (1993): 357–358.

Dionne, Deidra. "Are Some Countries Hacking the Paralympic System to Win More Medals?" November 15, 2016. https://www.cbc.ca/sports/ olympics/2.6605/paralympics-classification-system-1.3850121.

Docter, Shgufta, Moin Khan, Chetan Gohal, Bheeshma Ravi, Mohit Bhandari, Rajiv Gandhi, and Timothy Leroux. "Cannabis Use and Sport: A Systematic Review." *Sports Health* 12, no. 2 (2020): 189–199.

"Doctors Suggest Wrestling Ban." *Philadelphia Inquirer,* December 17, 1967, 13.

Doherty, Joseph. "From Ben Johnson to Barry Bonds: Sports' Steroid Scandal Continues." *Bleacher Report,* April 18, 2010. https://bleacherreport.com/ articles/381072-from-ben-johnson-to-barrybonds-sports-steroid-scan dal-continues.

Dolan, Eimear, Helen O'Connor, Adrian McGoldrick, Gillian O'Loughlin, Deirdre Lyons, and Giles Warrington. "Nutritional, Lifestyle, and Weight Control Practices of Professional Jockeys." *Journal of Sports Sciences* 29, no. 8 (2011): 791–799.

Donnelly, Peter. "Child Labour, Sport Labour: Applying Child Labour Laws to Sport." *International Review for the Sociology of Sport* 32, no. 4 (1997): 389–406.

"'Dope' Evil of the Turf." *New York Times*, October 19, 1903, 8.

Doscher, Nathan. "The Effect of Rapid Weight Loss upon the Performance of Wrestlers and Boxers and upon the Physical Proficiency of College Students." *Research Quarterly in Exercise and Sport* 15 (1944): 317–324.

Douglas, Thomas. "Enhancement in Sport, and Enhancement Outside Sport." *Studies in Ethics, Law, and Technology* 1, no. 1 (2009): ukpmcpa2293. doi: 10.2202/1941-6008.1000.

Doyle, Dave. "'Devastated' Kevin Lee Says Weight Cut 'Damn Near Killed Me.'" October 8, 2017. mmafighting.com/2017/10/8/16443192/devastated-kevin-lee-says-weight-cut-damn-near-killed-me.

DuBois, L. Zachary, and Heather Shattuck-Heidorn. "Challenging the Binary: Gender/sex and the Bio-logics of Normalcy." *American Journal of Human Biology* 33, no. 5 (2021): e23623.

Duncan, Jason. "A Re-Union for MMA: Reoccurring Issues Plaguing Mixed Marshal Arts Fighters and Potential Solutions." *University of Denver Sports and Entertainment Law Journal* 23 (2020): 11–44.

Dunning, Eric. "Sport as a Male Preserve: Notes on the Social Sources of Masculine Identity and Its Transformations." *Theory, Culture & Society* 3, no. 1 (1986): 79–90.

Dworkin, Gerald. "Paternalism." *The Monist*, 56 (1972): 64–84.

Ekblom, Björn, Alberto N. Goldbarg, and Bengt Gullbring. "Response to Exercise After Blood Loss and Reinfusion." *Journal of Applied Physiology* 33, no. 2 (1972): 175–180.

Ekfelt, Vernon. "Eliminating the Criticisms of High School Wrestling." *Athletic Journal* (December 1955): 10–11.

Elfenbaum, Louis. "The Physiological Effects of Rapid Weight Loss Among Wrestlers." PhD diss., Ohio State University, 1966.

Elsas, Louis J., Risa P. Hayes, and Kasinathan Muralidharan. "Gender Verification at the Centennial Olympic Games." *Journal of the Medical Association of Georgia* 86, no. 1 (1997): 50–54.

Elsas, Louis J., Arne Ljungqvist, Malcolm A. Ferguson-Smith, Joe Leigh Simpson, Myron Genel, Alison S. Carlson, Elizabeth Ferris, Albert De La

Chapelle, and Anke A. Ehrhardt. "Gender Verification of Female Athletes." *Genetics in Medicine* 2, no. 4 (2000): 249–254.

Epstein, David J. *The Sports Gene: Inside the Science of Extraordinary Athletic Performance.* New York: Penguin, 2014.

Erikainen, Sonja. "Hybrids, Hermaphrodites, and Sex Metamorphoses: Gendered Anxieties and Sex Testing in Elite Sport, 1937–1968." In *Gender Panic, Gender Policy,* edited by Vasilikie Demos and Marcia Texler Segal, 155–176. Leeds: Emerald Insight, 2017.

Erikainen, Sonja. "The Story of Mark Weston: Re-centering Histories and Conceptualising Gender Variance in 1930s International Sport." *Gender and History* 32, no. 2 (2020): 304–319.

Fainaru-Wada, Mark, and Lance Williams. *Game of Shadows: Barry Bonds, BALCO, and the Steroids Scandal That Rocked Professional Sports.* New York: Penguin, 2006.

Farstad, Siri. "Protecting Children's Rights in Sport: The Use of Minimum Age." *Human Rights Law Commentary* 3 (2007): 1–20.

Fausto-Sterling, Anne. *Sex/gender: Biology in a Social World.* New York: Routledge, 2012.

Fénichel, Patrick, Françoise Paris, Pascal Philibert, Sylvie Hiéronimus, Laura Gaspari, Jean-Yves Kurzenne, Patrick Chevallier, Stéphane Bermon, Nicolas Chevalier, and Charles Sultan. "Molecular Diagnosis of 5α-reductase Deficiency in 4 Elite Young Female Athletes through Hormonal Screening for Hyperandrogenism." *Journal of Clinical Endocrinology & Metabolism* 98, no. 6 (2013): E1055–E1059.

Ferrigon, Phillip, and Kevin Tucker. "Person-First Language vs. Identity-First Language: An Examination of the Gains and Drawbacks of Disability Language in Society." *Journal of Teaching Disability Studies* 1 (January 3, 2019). https://jtds.commons.gc.cuny.edu/person-first-language-vs-ident ity-first-language-an-examination-of-the-gains-and-drawbacks-of-dis ability-language-in-society/.

Foddy, Bennett, and Julian Savulescu. "Ethics of Performance Enhancement in Sport: Drugs and Gene Doping." *Principles of Health Care Ethics* (2006): 511–519.

Foddy, Bennett, and Julian Savulescu. "Time to Re-evaluate Gender Segregation in Athletics?" *British Journal of Sports Medicine* 45, no, 15 (2011): 1184–1188.

Foer, Franklin. "The Goals of Globalization." *Foreign Policy* 12, no. 5 (2005). https://foreignpolicy.com/2009/10/20/the-goals-of-globalization/.

Fotheringham, William. *Put Me Back on My Bike: In Search of Tom Simpson.* New York: Random House, 2012.

Franke, Werner W., and Brigitte Berendonk. "Hormonal Doping and Androgenization of Athletes: A Secret Program of the German Democratic Republic Government." *Clinical Chemistry* 43, no. 7 (1997): 1262–1279.

Franklin, Simon, Jonathan Ospina Betancurt, and Silvia Camporesi. "What Statistical Data of Observational Performance Can Tell Us and What They Cannot: The Case of Dutee Chand v. AFI & IAAF." *British Journal of Sports Medicine* 52, no. 7 (2018): 420–421.

Frost, Dennis J. *Seeing Stars: Sports Celebrity, Identity, and Body Culture in Modern Japan.* Cambridge, MA: Harvard University Press, 2010.

Frost, Norman. "Banning Drugs in Sports: A Skeptical View." *Hastings Center Report* 16, no. 4 (1986): 5–10.

Gal, Dubnov-Raz, Yael Mashiach-Arazi, Ariella Nouriel, Raanan Raz, and Naama W. Constantini, "Can Height Categories Replace Weight Categories in Striking Martial Arts Competitions? A Pilot Study." *Journal of Human Kinetics* 47 (2015): 91–98.

Gander, Kashmira. "Nutrigenomics: Can DNA be used to Change Your Fitness and Diet Regimen?" *Independent,* February 16, 2017. https://www.independent.co.uk/life-style/nutrigenomics-dna-fitness-diet-regime-dnafit-does-it-work-experts-genetic-gym-health-a7582966.html.

Gao, Guangping, Corinna Lebherz, Daniel J. Weiner, Rebecca Grant, Roberto Calcedo, Beth McCullough, Adam Bagg, Yi Zhang, and James M. Wilson. "Erythropoietin Gene Therapy Leads to Autoimmune Anemia in Macaques." *Blood* 103, no. 9 (2004): 3300–3302.

García, Ana De la Fuente. "Height Categories as a Healthier Alternative to Weight Categories in Taekwondo Competition." *Revista de Artes Marciales Asiáticas* 13, no. 1 (2018): 53–60.

Garcia, Marlen. "Attorney: Curry Won't Take DNA Test." *Chicago Tribune,* September 24, 2005, sec. 3, p. 1.

García, Raúl Sánchez, and Dominic Malcolm. "Decivilizing, Civilizing or Informalizing? The International Development of Mixed Martial Arts." *International Review for the Sociology of Sport* 45, no. 1 (2010): 39–58.

Gayagay, George, Bing Yu, Brett Hambly, Tanya Boston, Alan Hahn, David S. Celermajer, and Ronald J. Trent. "Elite Endurance Athletes and the ACE I Allele: The Role of Genes in Athletic Performance." *Human Genetics* 103, no. 1 (1998): 48–50.

Gibbs, Aimee E., Joel Pickerman, and Jon K. Sekiya. "Weight Management in Amateur Wrestling." *Sports Health* 1, no. 3 (2009): 227–230.

Gillum, Olden Curtice. "The Effects of Weight Reduction on the Bodily Strength of Wrestlers." PhD diss., Ohio State University, 1940.

Gisclair, S. Derby. *The Olympic Club of New Orleans: Epicenter of Professional Boxing, 1883–1897.* Jefferson, NC: McFarland, 2018.

Glasspiegel, Ryan. "Tara Lipinski Eviscerates New Figure Skating Age Limit: 'Broken System.'" *New York Post,* June 9, 2022. https://nypost.com/2022/06/09/tara-lipinski-eviscerates-new-figure-skating-age-limit/.

Gleaves, John. "Biometrics and Antidoping Enforcement in Professional Sport." *American Journal of Bioethics* 17, no. 1 (2017): 77–79.

Gleaves, John. "Doped Professionals and Clean Amateurs: Amateurism's Influence on the Modern Philosophy of Anti-Doping." *Journal of Sport History* 38, no. 2 (2011): 237–254.

Gleaves, John. "Enhancing the Odds: Horse Racing, Gambling and the First Anti-Doping Movement in Sport, 1889–1911." *Sport in History* 32, no. 1 (2012): 26–52.

Gleaves, John. "Manufactured Dope: How the 1984 US Olympic Cycling Team Rewrote the Rules on Drugs and Sports." *International Journal of the History of Sport* 32, no. 1 (2015): 89–107.

Goh, Chui Ling. "To What Extent Does the Paralympic Games Promote the Integration of Disabled Persons into Society?" *International Sports Law Journal* 20 (2020): 36–54.

Gold, John R. and Margaret M. Gold. "Access for All: The Rise of the Paralympic Games." *Perspectives in Public Health* 127, no. 3 (2007): 122–141.

Goodman, Susan. *Spirit of Stoke Mandeville: The Story of Sir Ludwig Guttmann.* London: HarperCollins, 1986.

Gorn, Elliott J. "The Bare-Knuckle Era." In *The Cambridge Companion to Boxing,* edited by Gerald Early, 34–51. Cambridge: Cambridge University Press, 2019.

Gorn, Elliott J. *The Manly Art: Bare-Knuckle Prize Fighting in America.* Ithaca, NY: Cornell University Press, 1986.

Grant, Paul. "'I'm Handing Back My Medal': Is Paralympic Sport Classification Fit for Purpose?" *BBC,* September 18, 2017. http://www.bbc.com/sport/disability-sport/41253174.

Gray, W. Russel. "For Whom the Bell Tolled, The Decline of British Prize Fight in the Victorian Era." *Journal of Popular Culture* 21, no. 2 (1987): 53–64.

Greenemeire, Larry. "How Olympians Could Beat the Competition by Tweaking Their Genes." *Smithsonian,* August 5, 2012. https://www.smithsonianmag.com/technology-space/how-olympians-could-beat-the-competition-by-tweaking-their-genes-14591201/#HLgF3Hj30OXYtGYV.99.

Greey, Ali Durham, and Helen Jefferson Lenskyj, eds. *Justice for Trans Athletes.* Leeds: Emerald Insight, 2023.

Grenfell, Christopher C., and Robert E. Rinehart. "Skating on Thin Ice: Human Rights in Youth Figure Skating." *International Review for the Sociology of Sport* 38, no. 1 (2003): 79–97.

Gruender, Scott A. *Jockey: The Rider's Life in American Thoroughbred Racing.* Jefferson, NC: McFarland and Company, 2007.

Guttmann, Allen. *From Ritual to Record: The Nature of Modern Sports.* New York: Columbia University Press, 1978.

Guttmann, Ludwig. "History of the National Spinal Injuries Centre, Stoke Mandeville Hospital, Aylesbury." *Proceeding of the Annual Scientific Meeting of the Society Held at Stoke Mandeville Hospital, Aylesbury,* July 27–29, 1967, 115–126.

Guttmann, Ludwig. "Looking Back on a Decade." *The Cord* 6, no. 4 (1954): 9–23.

Guttmann, Ludwig. "The Second National Stoke Mandeville Games for the Paralyzed." *The Cord* 3 (1949): 24.

Guttmann, Ludwig. *Textbook of Sport for the Disabled.* Aylesbury: HM & M Publishers, 1976.

Hagen, Thomas, Espen Tønnessen, and Stephen Seiler. "9.58 and 10.49: Nearing the Citius End for 100 m?" *International Journal of Sports Physiology and Performance* 10 (2015): 269–272.

Halabchi, Farzin. "Doping in Combat Sports." In *Combat Sports Medicine,* edited by Ramin Kordi, Nicol Maffulli, Randall R. Wroble, and W. Angus Wallace, 55–72. London: Springer, 2009.

Handelsman, David J., Angelica L. Hirschberg, and Stéphane Bermon. "Circulating Testosterone as the Hormonal Basis of Differences in Athletic Performance." *Endocrine Reviews* 39, no. 5 (2018): 803–829.

"Hannah Cockroft: 'Para-Classification Tests Are Humiliating.'" BBC, October 27, 2017. https://www.bbc.com/sport/disability-sport/41780947.

Hansen, Jørn. "The Origins of the Term Handicap in Games and Sports— History of a Concept." *Physical Culture and Sport Studies Research* 65 (2015): 7–13.

Hansen, Norman C. "Wrestling with "Making Weight." *Physician and Sportsmedicine* 6 (1978): 106–111.

Hanson, E. D., A. T. Ludlow, A. K. Sheaff, J. Park, and S. M. Roth. "ACTN3 Genotype Does Not Influence Muscle Power." *International Journal of Sports Medicine* 31, no. 11 (2010): 834–838.

Hanson, Sharon Kinney. *The Life of Helen Stephens: The Fulton Flash.* Carbondale: Southern Illinois University Press, 2004.

Hanstad, Dag Vidar, and Sigmund Loland. "Elite Athletes' Duty to Provide Information on Their Whereabouts: Justifiable Anti-Doping Work or an

Indefensible Surveillance Regime?" *European Journal of Sport Science* 9, no. 1 (2009): 3–10.

Hanstad, Dag Vidar, Andy Smith, and Ivan Waddington. "The Establishment of the World Anti-Doping Agency: A Study of the Management of Organizational Change and Unplanned Outcomes." *International Review for the Sociology of Sport* 43, no. 3 (September 2008): 227–249.

Harmon, Kimberly G., Jonathan A. Drezner, David Klossner, and Irfan M. Asif. "Sickle Cell Trait Associated with a RR of Death 37 Times in National Collegiate Athletic Association Football Athletes: A Database with 2 Million Athlete-Years as the Denominator." *British Journal of Sports Medicine* 46, no. 5 (2012): 325–330.

Harper, Joanna, Maria-Jose Martinez-Patino, Fabio Pigozzi, and Yannis Pitsiladis. "Implications of a Third Gender for Elite Sports." *Current Sports Medicine Reports* 17, no. 2 (2018): 42–44.

Hay, Eduardo. "Sex Determination in Putative Female Athletes." *JAMA* 221, no. 9 (1972): 998–999.

Healy, Marie-Louise, James Gibney, Claire Pentecost, Mike J. Wheeler, and Peter H. Sönksen. "Endocrine Profiles in 693 Elite Athletes in the Postcompetition Setting." *Clinical Endocrinology* 81, no. 2 (2014): 294–305.

Heerdt, Daniela. "The Court of Arbitration for Sport: Where Do Human Rights Stand?" Centre for Sport and Human Rights, May 10, 2019. https://sporthumanrights.org/library/the-court-of-arbitration-for-sport-where-do-human-rights-stand/.

Heggie, Vanessa. "Subjective Sex: Science, Medicine and Sex Tests in Sports." In *Transgender Athletes in Competitive Sport*, edited by Eric Anderson and Ann Travers, 131–142. London: Taylor & Francis, 2017.

Heggie, Vanessa. "Testing Sex and Gender in Sports; Reinventing, Reimagining and Reconstructing Histories." *Endeavour* 34, no. 4 (2010): 157–163.

Heineman, David S. "Leaving Fandom: Why I Gave Up Sports, Why You Should Consider It, and How to Start." *Medium.* September 21, 2016. https://medium.com/@DrHeineman/leaving-fandom-why-i-gave-up-sports-why-you-should-consider-it-and-how-to-start-6d1941c1a915.

Heinonen, Janet. "A Decent Proposal: Keeping Track." *International Track and Field Newsletter,* March 1994, 24.

"Helen Stephens A Man, Polish Writer Thinks." *Chicago Tribune,* August 6, 1936, 20.

Henne, Kathryn, and Madeleine Pape. "Dilemmas of Gender and Global Sports Governance: An Invitation to Southern Theory." *Sociology of Sport Journal* 35, no. 3 (2018): 216–225.

Henson, Stanley W. "The Problem of Losing Weight," *Amateur Wrestling News,* February 12, 1969, 14.

Hersh, Philip. "Are Ski Jumpers Too Thin?" *Chicago Tribune,* January 16, 2002. http://articles.chicagotribune.com/2002-01-16/sports/020116039 7_1_top-jumpers-alan-alborn-sven-hannawald.

Hersh, Philip. "Youth, Maturity—Gymnastics Needs Best of Both Worlds." *Chicago Tribune,* August 24, 1995. https://www.chicagotribune.com/ news/ct-xpm-1995-08-24-9508240182-story.html.

Heuberger, Jules A. A. C., and Adam F. Cohen. "Review of WADA Prohibited Substances: Limited Evidence for Performance-Enhancing Effects." *Sports Medicine* 49, no. 4 (2019): 525–539.

Hines, James R. *Figure Skating: A History*. Urbana: University of Illinois Press, 2006.

Hirschberg, Angelica Lindén. "Hyperandrogenism in Female Athletes." *Journal of Clinical Endocrinology and Metabolism* 104, no. 2 (2019): 503–505.

Hoad, Neville. "Run, Caster Semenya, Run! Nativism and the Translations of Gender Variance." *Safundi* 11, no. 4 (2010): 397–405.

Hoberman, John. "How Drug Testing Fails: The Politics of Doping Control." In *Doping in Elite Sport: Politics of Drugs in the Olympic Movement*, edited by Wayne Wilson and Edard Derse, 241–274. Champaign, IL: Human Kinetics, 2001.

Hoberman, John. *Mortal Engines: The Science of Performance and the Dehumanization of Sport*. New York: Free Press, 1992.

Holm, Søren. "Doping under Medical Control—Conceptually Possible but Impossible in the World of Professional Sports?" *Sports, Ethics and Philosophy* 1, no. 2 (2007): 135–145.

Holmes, Tracey. "Allegations of Cheating, Threats and Cover-ups Aimed at Australian Paralympic Swimming." *ABC News,* December 2, 2017. http:// www.abc.net.au/news/2017-12-03/allegations-target-australian-paralym pic-swimming/9221084.

Holzer, Lena. "What Does It Mean to Be a Woman in Sports? An Analysis of the Jurisprudence of the Court of Arbitration for Sport." *Human Rights Law Review* 20, no. 3 (2020): 387–411.

Houlihan, Barrie, Dag Vidar Hanstad, Sigmund Loland, and Ivan Waddington. "The World Anti-Doping Agency at 20: Progress and Challenges." *International Journal of Sport Policy and Politics* 11, no. 2 (2019): 193–201.

Houston, Michael. "Norway Submit Figure Skating Age Limit Rise Despite Backlash." *Inside the Games,* November 30, 2020. https://www.insidethega mes.biz/articles/1101434/norway-figure-skating-age-limit.

Howe, P. David. "From Inside the Newsroom: Paralympic Media and the 'Production' of Elite Disability." *International Review for the Sociology of Sport* 43, no. 2 (2008): 135–150.

Howe, P. David, and Carwyn Jones. "Classification of Disabled Athletes:(Dis) empowering the Paralympic Practice Community." *Sociology of Sport Journal* 23, no. 1 (2006): 29–46.

Huang, Grace, and Shehzad Basaria. "Do Anabolic-androgenic Steroids Have Performance-enhancing Effects in Female Athletes?" *Molecular and Cellular Endocrinology* 464 (2018): 56–64.

Huggins, Mike. "Racing Culture, Betting, and Sporting Protomodernity: The 1975 Newmarket Carriage Match." *Journal of Sport History* 42, no. 2 (2015): 322–339.

Human Rights Campaign. "Glossary of Terms." Accessed December 21, 2021. https://www.hrc.org/resources/glossary-of-terms.

Human Rights Watch. "End Abusive Sex Testing for Women Athletes." December 4, 2020. https://www.hrw.org/news/2020/12/04/end-abus ive-sex-testing-women-athletes#.

Human Rights Watch. *They're Chasing Us Away from Sport: Human Rights Violations in Sex Testing of Elite Athletes*, December 4, 2020. https://www. hrw.org/report/2020/12/04/theyre-chasing-us-away-sport/human-rig hts-violations-sex-testing-elite-women#.

Hunt, Thomas M. *Drug Games: The International Olympic Committee and the Politics of Doping, 1960–2008.* Austin: University of Texas Press, 2011.

Hunt, Thomas M., Paul Dimeo, Florian Hemme, and Anne Mueller. "The Health Risks of Doping during the Cold War: A Comparative Analysis of the Two Sides of the Iron Curtain." *International Journal of the History of Sport* 31, no. 17 (2014): 2230–2244.

Ialongo, Nicola, Raphael Hermann, and Lorenz Rahmstorf, "Bronze Age Weight Systems as a Measure of Market Integration in Western Eurasia." *PNAS* 118, no. 27 (2021): e2105873118.

"Indian Runner Fails Gender Test, Loses Medal." ESPN.com. December 18, 2006. http://sports.espn.go.com.

Ingle, Sean. "'Tremendous Coldness': IOC President Condemns Kamila Valieva's Entourage." *The Guardian,* February 18, 2022. https://www.theg uardian.com/sport/2022/feb/18/tremendous-coldness-ioc-president- slams-kamila-valievas-entourage-over-skaters-treatment.

"The Inside Story of a Toxic Culture at Maryland Football." *ESPN.* August 10, 2018. https://www.espn.com/college-football/story/_/id/24342005/ maryland-terrapins-football-culture-toxic-coach-dj-durkin.

International Amateur Athletic Federation. *Handbook of the International Amateur Athletic Federation, 1927–1928.* Accessed December 22, 2023. https://www.iaaf.org/news/news/a-piece-of-anti-doping-history-iaaf-handbook.

International Association of Athletics Federation. "Caster Semenya May Compete." July 6, 2010. https://worldathletics.org/news/iaaf-news/caster-semenya-may-compete.

International Association of Athletics Federations. "Decisions Made at IAAF Council Meetings in Doha." October 14, 2019. https://www.worldathletics.org/news/press-release/iaaf-council-219-decisions.

International Association of Athletics Federations. *IAAF Competition Rules, 2008.* https://www.worldathletics.org/download/download?filename= 5a5b59ac-eb58-45e4-a9a3-003c1e61d619.pdf&urlslug=iaaf%20track%20 and%20field%20facilities%20manual%202008%20edition%20-%20chapt ers%201-3.

International Association of Athletics Federations. *IAAF Regulations Governing Eligibility of Females with Hyperandrogenism to Compete in Women's Competition.* 2011. https://www.worldathletics.org/news/iaaf-news/iaaf-to-introduce-eligibility-rules-for-femal-1.

International Association of Athletics Federations. "Oscar Pistorius— Independent Scientific Study Concludes That Cheetah Prosthetics Offer Clear Mechanical Advantages." January 14, 2008. https://www.worldat hletics.org/news/news/oscar-pistorius-independent-scientific-stud-1.

International Federation for Athletes with Intellectual Impairments. "Athlete Eligibility." https://inas.org/about-us/athlete-eligibility/eligibility-and-classification.

International Olympic Committee. Annual Report, 2021. https://stillmed. olympics.com/media/Documents/International-Olympic-Committee/ Annual-report/IOC-Annual-Report-2021.pdf?_ga=2.233961374.123 1423006.1662651976-399267994.1662651976.

International Olympic Committee. IOC Consensus Meeting on Sex Reassignment and Hyperandrogenism. November 2015. https://stillmed. olympic.org/Documents/Commissions_PDFfiles/Medical_commission/ 2015-11_ioc_consensus_meeting_on_sex_reassignment_and_hyperandr ogenism-en.pdf.

International Olympic Committee. IOC Framework on Fairness, Inclusion and Non-Discrimination on the Basis of Gender Identity and Sex Variations. 2022. https://stillmed.olympics.com/media/Documents/ News/2021/11/IOC-Framework-Fairness-Inclusion-Non-discriminat

ion-2021.pdf?_ga=2.116948229.2094909257.1637082260-499116176.163
4933505.

International Olympic Committee. IOC Regulations on Female
Hyperandrogenism. 2012. https://stillmed.olympic.org/Documents/
Commissions_PDFfiles/Medical_commission/2012-06-22-IOC-Regu
lations-on-Female-Hyperandrogenism-eng.pdf.

International Olympic Committee. *Olympic Charter*. Lausanne: International
Olympic Committee, 2021.

International Olympic Committee. Statement of the Stockholm Consensus
on Sex Reassignment in Sports. 2003. https://stillmed.olympic.org/
Documents/Reports/EN/en_report_905.pdf.

International Olympic Committee. "Tokyo 2020 Event Programme To See
Major Boost for Female Participation, Youth and Urban Appeal." June
9, 2017. https://www.olympic.org/news/tokyo-2020-event-programme-
to-see-major-boost-for-female-participation-youth-and-urban-appeal.

International Paralympic Committee. "Guide to Reporting on Persons with
an Impairment." October 2014. https://m.paralympic.org/sites/default/
files/document/141027103527844_2014_10_31+Guide+to+reporting+
on+persons+with+an+impairment.pdf.

International Paralympic Committee. *IPC Athlete Classification Code*. Bonn:
International Paralympic Committee, 2015. https://www.paralympic.org/
sites/default/files/2020-05/170704160235698_2015_12_17%2BClassificat
ion%2BCode_FINAL2_0-1.pdf

International Paralympic Committee. *IPC Classification Code: Models of Best
Practice, Intentional Misrepresentation Rules*. Bonn: International Paralympic
Committee, 2013. https://oldwebsite.paralympic.org/sites/default/files/
document/141113161802225_2014_10_10+Sec+ii+chapter+1_3_Mod
els+of+best+practice_+Intentional+Misrepresentation+Rules.pdf

International Paralympic Committee. "IPC Statement on USA Swimmer
Victoria Arlen." August 12, 2013. https://www.paralympic.org/news/ipc-
statement-usa-swimmer-victoria-arlen.

International Paralympic Committee. IPC Swimming Classification Rules
and Regulations. 2011. Accessed March 11, 2020. https://www.paralym
pic.org/sites/default/files/document/120706163426076_2011_05_30__
Swimming_Classification_Regulations.pdf

International Paralympic Committee. *Models of Best Practice: National
Classification*. Bonn: International Paralympic Committee, 2017. https://
www.paralympic.org/sites/default/files/document/170216081042262_
2017_02_16+Models+of+Best+Practice_National+Classification.pdf.

International Paralympic Committee. "What Is Classification." Accessed December 22, 2023. https://www.paralympic.org/classification.

International Skating Union. *Agenda of the 58th Ordinary Congress, Phuket, 2022.* Accessed December 22, 2023. https://www.isu.org/docman-documents-links/isu-files/documents-communications/about-isu/congress-documents/28312-isu-communication-2472-1/file.

International Skating Union. "#FigureSkating." February 22, 2023. https://www.isu.org/isu-news/news/145-news/14500-isu-appeals-case-of-roc-figure-skater-to-court-of-arbitration-for-sport?templateParam=15.

International Ski Federation, Rule 1.2.1.1. "Ski Length." In *Specifications for Competition Equipment, Edition 2018/2019.* Oberhofen, Switzerland: International Ski Federation, 2018.

Irgan, Umair, and Julia Bellus. "Olympic Swimmer Ryan Lochte Broke Doping Rules. It Happens Far More than You Think." *Vox,* July 27, 2018. https://www.vox.com/2018/7/24/17603358/ryan-lochte-doping-ban-olympics-instagram.

Janofsky, Michael, and Peter Alfano. "Drug Use by Athlete Runs Free Despite Tests." *New York Times,* November 17, 1988. https://www.nytimes.com/1988/11/17/sports/drug-use-by-athletes-runs-free-despite-tests.html.

Jederström, Moa, Sara Agnafors, Christina Ekegren, Kristina Fagher, Håkan Gauffin, Laura Korhonen, Jennifer Park, Armin Spreco, and Toomas Timpka. "468 Determinants of Sports Injury in Young Female Swedish Competitive Figure Skaters." *British Journal of Sports Medicine* 55, Suppl 1 (2021): A179–A179.

Jenkins, Sally. "There's a Legal Remedy to the Doping Issue." *Washington Post,* October 12, 2007.

Jetton, Adam M., Marcus M. Lawrence, Marco Meucci, Tracie L. Haines, Scott R. Collier, David M. Morris, and Alan C. Utter. "Dehydration and Acute Weight Gain in Mixed Martial Arts Fighters Before Competition." *Journal of Strength and Conditioning Research* 27, no. 5 (2013): 1322–1326.

John, Joseph, Gretchen Kerr, and Simon Darnell. "'Safe Sport Is Not for Everyone': Equity-deserving Athletes' Perspectives of, Experiences, and Recommendations for Safe Sport." *Frontiers in Psychology* 13 (2022). https://doi.org/10.3389/fpsyg.2022.832560

Jones, Bethany Alice, Jon Arcelus, Walter Pierre Bouman, and Emma Haycraft. "Sport and Transgender People: A Systematic Review of the Literature Relating to Sport Participation and Competitive Sport Policies." *Sports Medicine* 47, no. 4 (2017): 701–716.

Jordan-Young, Rebecca, and Katrina Karkazis. *Testosterone: An Unauthorized Biography.* Cambridge, MA: Harvard University Press, 2019.

Juengst, Eric T. "Subhuman, Superhuman, and Inhuman: Human Nature and the Enhanced Athlete." In *Athletic Enhancement, Human Nature, and Ethics*, edited by Jan Tolleneer, Sigrid Sterckx, Pieter Bonte, 89–103. Dordrecht: Springer, 2013.

"'Junction Boys Syndrome': How College Football Fatalities Became Normalized." *The Guardian,* August 19, 2018. https://www.theguardian.com/sport/2018/aug/19/college-football-deaths-offseason-workouts.

Kane, Mary Jo. "Resistance/transformation of the Oppositional Binary: Exposing Sport as a Continuum." *Journal of Sport and Social Issues* 19, no. 2 (1995): 191–218.

Karkazis, Katrina. "The Misuses of 'Biological Sex.'" *The Lancet* 394, no. 10212 (2019): 1898–1899.

Karkazis, Katrina, and Rebecca M. Jordan-Young. "The Powers of Testosterone: Obscuring Race and Regional Bias in the Regulation of Women Athletes." *Feminist Formations* 30, no. 2 (2018): 1–39.

Karkazis, Katrina, Rebecca Jordan-Young, Georgiann Davis, and Silvia Camporesi. "Out of Bounds? A Critique of the New Policies on Hyperandrogenism in Elite Female Athletes." *American Journal of Bioethics* 12, no. 7 (2012): 3–16.

Kass, Leon R. *Beyond Therapy: Biotechnology and the Pursuit of Happiness, A Report of The President's Council on Bioethics.* New York: Dana Press, 2003.

Kayser, Bengt, and Barbara Broers. "Doping and Performance Enhancement: Harms and Harm Reduction." In *Routledge Handbook of Drugs and Sport*, edited by Verner Møller, Ivan Waddington, and John Hoberman, 363–376. London: Routledge, 2015.

Kayser, Bengt, Alexandre Mauron, and Andy Miah. "Current Ant-Doping Policy: A Critical Appraisal." *BMC Medical Ethics* 8, no. 2 (2007). https://bmcmedethics.biomedcentral.com/articles/10.1186/1472-6939-8-2.

Kayser, Bengt, and Jan Tolleneer. "Ethics of a Relaxed Antidoping Rule Accompanied by Harm-Reduction Measures." *Journal of Medical Ethics* 43 (2017): 282–286.

Kelly, Mike. "Paralympic Cheats Have Cost Me Medals Says Cramlington Legend Stephen Miller." *Chronicle,* November 7, 2017. https://www.chroniclelive.co.uk/news/north-east-news/paralympic-cheats-cost-medals-says-13864613.

Kenney, H. E. "The Problem of Making Weight for Wrestling Meets." *Journal of Health and Physical Education* 1 (1930): 24–25; 49.

Kerr, Roslyn, and C. Obel. "Reassembling Sex: Reconsidering Sex Segregation Policies in Sport." *International Journal of Sport Policy and Politics* 10, no. 2 (2018): 305–320.

Khodaee, Morteza, Lucianne Olewinski, Babk Shadgan, and Robert R. Kiningham. "Rapid Weight Loss in Sports with Weight Classes." *Current Sports Medicine Reports* 14, no. 6 (2015): 435–441.

Kidd, Bruce, and Peter Donnelly. "Human Rights in Sports." *International Review for the Sociology of Sport* 35, no. 2 (2000): 131–148.

Kidd, Bruce, Robert Edelman, and Susan Brownell. "Comparative Analysis of Doping Scandals: Canada, Russia, and China." In *Doping in Elite Sport: The Politics of Drugs in the Olympic Movement*, edited by Wayne Wilson and Ed Derse, 153–188. Champaign, IL: Human Kinetics, 2001.

Kirkwood, Ken. "Considering Harm Reduction as the Future of Doping Control Policy in International Sport." *Quest* 61, no. 2 (2009): 180–190.

Knapp, Gwen. "Why Ukraine's Small Paralympic Team Packs Such a Big Punch." *New York Times,* September 6, 2021.

Knox, Taryn, Lynley C. Anderson, and Alison Heather. "Transwomen in Elite Sport: Scientific and Ethical Considerations." *Journal of Medical Ethics* 45, no. 6 (2019): 395–403.

Kopp, J. David. "Eye on the Ball: An Interview with Dr. C. Stephen Johnson and Mark McGwire." *Journal of the American Optometric Association* 70, no. 2 (1999): 79–84.

Kordi, Ramin, Nicol Maffulli, Randall R. Wroble, and W. Angus Wallace, eds. *Combat Sports Medicine*. London: Springer, 2009.

Kowalczyk, Agnieszka D., Ellen T. Geminiani, Bridge W. Dahlberg, Lyle J. Micheli, and Dai Sugimoto. "Pediatric and Adolescent Figure Skating Injuries: A 15-year Retrospective Review." *Clinical Journal of Sport Medicine* 31, no. 3 (2019): 295–303.

Krane, Vikki, ed. *Sex, Gender, and Sexuality in Sport: Queer Inquiries*. London: Routledge, 2019.

Krane, Vikki, and Heather Barber. "Creating A New Sport Culture: Reflections on Queering Sport." In *Sex, Gender, and Sexuality in Sport: Queer Inquiries*, edited by Vikki Krane, 223–237. London: Routledge, 2019.

Krieger, Jörg, Lindsay Parks Pieper, and Ian Ritchie. "Sex, Drugs and Science: The IOC's and IAAF's Attempts to Control Fairness in Sport." *Sport in Society* 22, no. 9 (2019): 1555–1573.

Kyle, Donald G. "Greek Athletic Competitions: The Ancient Olympics and More." In *A Companion to Sport and Spectacle in Greek and Roman Antiquity*, edited by Paul Christesen and Donald G. Kyle, 17–35. Chichester: Wiley Blackwell, 2013.

Labadarios, Demete, Juan Kotze, D. Momberg, and T. V. W. Kotze. "Jockeys and Their Practices in South Africa." *Nutrition and Fitness for Athletes* 71 (1993): 97–114.

Labanowich, Stan, and Armand "Tip" Thiboutot. *Wheelchairs Can Jump! A History of Wheelchair Basketball*. Boston: Acanthus, 2011.

Larned, Deborah. "The Femininity Test: A Woman's First Olympic Hurdle." *Womensports*, July 1976, 8–11; 41.

Lasagna, Louis. "The Pharmaceutical Revolution: Its Impact on Science and Society." *Science* 166, no. 3910 (1969): 1227–1233.

Lawrence, D. W. "Sociodemographic Profile of an Olympic Team." *Public Health* 148 (2017): 149–158.

Lazarou, Stephen, Louis Reyes-Vallejo, and Abraham Morgentaler. "Wide Variability in Laboratory Reference Values for Serum Testosterone." *Journal of Sex Medicine* 3, no. 6 (2006): 1085–1089.

Le Clair, Jill M., ed. *Disability in the Global Sport Arena: A Sporting Chance*. London: Routledge, 2012.

Le Page, Michael. "Gene Doping in Sport Could Make the Olympics Fairer and Safer." *New Scientist,* August 5, 2016. https://www.newscientist.com/article/2100181-gene-doping-in-sport-could-make-the-olympics-fairer-and-safer/#ixzz7C2BGRZJO.

Legg, David. "Paralympic Games: History and Legacy of a Global Movement." *Physical Medicine Rehabilitation Clinics of North America* 29, no. 2 (2018): 417–425.

Legg, David, and Robert Steadward. "The Paralympic Games and 60 Years of Change (1948–2008): Unification and Restructuring from a Disability and Medical Model to Sport-Based Competition." In *Disability in the Global Sport Arena: A Sporting Chance*, edited by Jill M. Le Clair, 26–30. London: Routledge, 2012.

Legg, Kylie, Darryl Cochrane, Erica Gee, and Chris Rogers. "Jockey Career Length and Risk Factors for Loss from Thoroughbred Race Riding." *Sustainability* 12, no. 18 (2020): 7443.

Lemon, Jason. "China Will Begin Using Genetic Testing to Select Olympic Athletes." *Newsweek,* August 31, 2018. https://www.newsweek.com/china-begin-using-genetic-testing-select-olympic-athletes-1099058.

Lenskyj, Helen Jefferson. *Gender, Athletes' Rights, and the Court of Arbitration for Sport*. Leeds: Emerald Insight, 2018.

Lenskyj, Helen Jefferson. *The Olympic Games: A Critical Approach*. Leeds: Emerald Insight, 2020.

Levine, Benjamin D. "Should 'Artificial' High Altitude Environments Be Considered Doping?" *Scandinavian Journal of Medicine and Science in Sports* 16 (2006): 297–301.

Leydon, Mark A., and Clare Wall. "New Zealand Jockeys' Dietary Habits and Their Potential Impact on Health." *International Journal of Sport Nutrition and Exercise Metabolism* 12, no. 2 (2002): 220–237.

Lightfoot, J. Timothy, Monica J. Hubal, and Stephen M. Roth, eds. *Routledge Handbook of Sport and Exercise Systems Genetics*. New York: Routledge, 2019.

Limón, Iliana. "Sickle Cell Trait: The Silent Killer." *Orlando Sentinel*, July 24, 2011, C1.

Lipetz, Jennifer, and Roger J. Kruse. "Injuries and Special Concerns of Female Figure Skaters." *Clinics in Sports Medicine* 19, no. 2 (2000): 369–380.

Lippi, Giuseppe, Giuseppe Banfi, Emmanuel J. Favaloro, Joern Rittweger, and Nicola Muffalli. "Updates on Improvement of Human Athletic Performance: Focus on World Records in Athletics." *British Medical Bulletin* 87 (2008): 7–15.

Lippi, Guiseppe, Emmanuel J. Favaloro, and Gian Cesare Guidi. "The Genetic Basis of Human Athletic Performance: Why Are Psychological Components So Often Overlooked?" *Journal of Physiology* 586, no. 12 (2008): 3817.

Lippi, Giuseppe, Massimo Franchini, and Gian Cesare Guidi. "Prohibition of Artificial Hypoxic Environments in Sports: Health Risks Rather than Ethics." *Applied Physiology, Nutrition, and Metabolism* 32, no. 6 (2007): 1206–1207.

Ljungqvist, Arne, and Myron Genel. "Transsexual Athletes—When Is Competition Fair?" *Lancet* 366 (2005): S42–S43.

Ljungqvist, Arne, and Joe Leigh Simpson. "Medical Examination for Health of All Athletes Replacing the Need for Gender Verification in International Sports: The International Amateur Athletic Federation Plan." *JAMA* 267, no. 6 (1992): 850–852.

Loland, Sigmund, and Michael J. McNamee. "The 'Spirit of Sport,' WADAs Code Review, and the Search for An Overlapping Consensus." *International Journal of Sport Policy and Politics* 11, no. 2 (2019): 325–339.

Longman, Jeré. "Battle of Weight Versus Gain in Ski Jumping." *New York Times*, February 11, 2010. https://www.nytimes.com/2010/02/12/sports/olympics/12skijump.html?mtrref=www.google.com&gwh=6C5ADE01DF8E74AEA0C123D0BFC8CB3E&gwt=pay.

Longrigg, Roger. *The History of Horse Racing*. New York: Stein and Day, 1972.

Loosemore, Sandra. "'Figures' Don't Add Up in Competition Anymore." *CBS SportsLine*, December 16, 1998. https://web.archive.org/web/20080727021537/http://cbs.sportsline.com/u/women/skating/dec98/loosemore121698.htm.

López, Bernat. "Creating Fear: The 'Doping Deaths,' Risk Communication and the Anti-Doping Campaign." *International Journal of Sport Policy and Politics*, 6, no. 2 (2014): 213–225.

Los Angeles Olympic Organization Committee. Games of the XXIIIrd Olympiad, Los Angeles 1984. *International Olympic Medical Controls Brochure*, 1980. http://www.la84foundation.org.

Lucas, Charles J. P. *The Olympic Games, 1904.* St. Louis, MO: Woodward and Tiernan, 1905.

Lukaski, Henry C., ed. *Body Composition: Health and Performance in Exercise and Sport.* Boca Raton, FL: CRC Press, 2017.

MacDonald, James. "MMA: The Dangers of Cutting Weight in Mixed Martial Arts." *Bleacher Report,* January 16, 2013. https://bleacherreport.com/articles/1487089-mma-the-dangers-of-cutting-weight-in-mixed-martial-arts.

Macedo, Emmanuel, Matt Englar-Carlson, Tim Lehrbach, and John Gleaves. "Moral Communities in Anti-Doping Policy: A Response to Bowers and Paternoster." *Sport, Ethics and Philosophy* 13, no. 1 (2019): 49–61.

Mackay, Duncan. "The Dirtiest Race in History: Olympic 100m Final, 1988." *The Guardian,* April 17, 2003. https://www.theguardian.com/sport/2003/apr/18/athletics.comment.

Mackay, Duncan. "Tony Banks Criticises IOC at the World Conference on Doping." *The Guardian,* February 3, 1999. https://www.theguardian.com/sport/1999/feb/03/tony-banks-criticises-ioc-conference-doping-sport.

Macur, Juliet. "A Spectacle That Shook the World of Skating." *New York Times,* February 19, 2022.

Maffetone Philip B., and Paul B. Laursen. "Athletes: Fit but Unhealthy?" *Sports Medicine—Open* 2, no. 1 (2016), 24.

Magubane, Zine. "Spectacles and Scholarship: Caster Semenya, Intersex Studies, and the Problem of Race in Feminist Theory." *Signs* 39, no. 3 (2014): 761–785.

Marcellini, Anne, Sylvain Ferez, Damien Issanchou, Eric De Léséleuc, and Mike McNamee. "Challenging Human and Sporting Boundaries: The Case of Oscar Pistorius." *Performance Enhancement & Health* 1, no. 1 (2012): 3–9.

Marren, Amy. Facebook Post, February 1, 2020. https://www.facebook.com/amy.marren/posts/10222643039731150.

Martensen, Carsten Kraushaar, and Verner Møller. "More Money—Better Anti-Doping?" *Drugs: Education, Prevention and Policy* 24, no. 3 (2017): 286–294.

Martínez-Patiño, María José. "Personal Account: A Woman Tried and Tested." *Lancet* 366 (December 2005): S38.

Martínková, Irena. "Unisex Sports: Challenging the Binary." *Journal of the Philosophy of Sport* 47, no. 2 (2020): 248–265.

Maschke, Karen J. "Performance-Enhancing Technologies and the Ethics of Human Subjects Research." In *Performance-Enhancing Technologies in Sports: Ethical, Conceptual, and Scientific Issues*, edited by Thomas H. Murray, Karen J. Maschke, and Angela A. Wasunna, 97–110. Baltimore: Johns Hopkins University Press, 2009.

Massidda, Myosotis, Valeria Bachis, Laura Corrias, Francesco Piras, Marco Scorcu, Claudia Culigioni, Daniele Masala, and Carla M. Calò. "ACTN3 R577X Polymorphism Is Not Associated with Team Sport Athletic Status in Italians." *Sports Medicine* 1, no. 1 (2015): 1–5.

Mataruna-Dos-Santos, Leonardo José, Andressa Fontes Guimarães-Mataruna, and Daniel Range. "Paralympians Competing in the Olympic Games and the Potential Implications for the Paralympic Games." *Brazilian Journal of Education, Technology, and Society* 11, no. 1 (2018): 105–116.

Matthews, Joseph John, and Ceri Nicholas. "Extreme Rapid Weight Loss and Rapid Weight Gain Observed in UK Mixed Martial Arts Athletes Preparing for Competition." *International Journal of Sport Nutrition and Exercise Metabolism* 27, no. 2 (2017): 122–129.

McAuley, Alexander B. T., Joseph Baker, and Adam L. Kelly. "Defining 'Elite' Status in Sport: From Chaos to Clarity." *German Journal of Exercise and Sport Research* 52 (2022): 183–197.

McCanliss, Irene. *Weight on the Thoroughbred Horse*. Chester, MA, privately printed, 1967.

McClearen, Jennifer. *Fighting Visibility: Sports Media and Female Athletes in the UFC*. Urbana: University of Illinois Press, 2021.

McDonagh, Eileen, and Laura Pappano. *Playing with the Boys: Why Separate is Not Equal in Sports*. New York: Oxford University Press, 2009.

McDonald, Mary G. "Screening Saviors? The Politics of Care, College Sports, and Screening Athletes for Sickle Cell Trait." In *Sports, Society, and Technology: Bodies, Practices, and Knowledge Production*, edited by Jennifer J. Sterling and Mary G. McDonald, 247–267. Gate East, Singapore: Palgrave Macmillan, 2020.

McGeehan, Nicholas. "Spinning Slavery: The Role of the United States and UNICEF in the Denial of Justice for the Child Camel Jockeys of the United Arab Emirates." *Journal of Human Rights Practice* 5, no. 1 (2013): 96–124.

McNamee, Mike. "Doping Scandals, Rio, and the Future of Anti-Doping Ethics." *Sport, Ethics and Philosophy* 10, no. 2 (2016): 113–116.

McNamee, Michael J. "The Spirit of Sport and Anti-Doping Policy: An Ideal Worth Fighting For." *Play True* 1 (2013): 14–16.

McNamee, Michael J. "The Spirit of Sport and the Medicalisation of Anti-Doping: Empirical and Normative Ethics." *Asian Bioethics Review* 4, no. 4 (2012): 374–392.

McNamee, Mike, and Verner Møller, eds. *Doping and Anti-Doping Policy in Sport: Ethical, Legal and Social Perspectives.* London: Taylor & Francis, 2011.

McPherron, Alexandra C., and Se-Jin Lee. "Double Muscling in Cattle Due to Mutations in the Myostatin Gene." *Proceedings of the National Academy of Sciences* 94, no. 23 (1997): 12457–12461.

"Medical Commission." *Olympic Newsletter* 5 (1968): 71–73.

Menier, Amanda. "Use of Event-Specific Tertiles to Analyse the Relationship between Serum Androgens and Athletic Performance." *British Journal of Sports Medicine* 52, no. 23 (2018): 1540.

Messner, Michael A. "Sports and Male Domination: The Female Athlete as Contested Ideological Terrain." *Sociology of Sport Journal* 5, no. 3 (1988): 197–211.

Messner, Michael A. *Taking the Field: Women, Men, and Sports.* Minneapolis: University of Minnesota Press, 2002.

Miah, Andy. *Genetically Modified Athletes: Biomedical Ethics, Gene Doping and Sport.* London: Routledge, 2004.

Miah, Andy. "Rethinking Enhancement in Sport." *Annals New York Academy of Sciences* 1093, no. 1 (2006): 301–320.

Miah, Andy, and Emma Rich. "Genetic Tests for Ability? Talent Identification and the Value of an Open Future." *Sport, Education and Society* 11, no. 3 (2006): 259–273.

Middleton, Iris M. "Cockfighting in Yorkshire during the Early Eighteenth Century." *Northern History* 40, no. 1 (2003): 129–146.

Mignon, Patrick. "The Tour de France and the Doping Issue." *International Journal of the History of Sport* 20, no. 2 (2003): 227–245.

Miller, Fiona Alice. "'Your True and Proper Gender': The Barr Body as a Good Enough Science of Sex." *Studies in History and Philosophy of Science Part C: Studies in History and Philosophy of Biological and Biomedical Sciences* 37, no. 3 (2006): 459–483.

Minsberg, Talya. "In Her New Book, Kara Goucher Keeps Running Accountable." *New York Times,* March 25, 2023. https://www.nytimes.com/2023/03/25/sports/running-goucher-longest-race.html.

Mitra, Payoshni. "The Untold Stories of Female Athletes with Intersex Variations in India." In *Routledge Handbook of Sport, Gender and Sexuality*, edited by Eric Anderson and Jennifer Hargreaves, 384–394. London: Routledge, 2014.

Moeslein, Anna. "Watch Kamila Valieva Become the First Woman to Land a Quad at the Olympics." *Glamour,* February 7, 2022. https://www.glamour.com/story/watch-kamila-valieva-become-the-first-woman-to-land-a-quad-at-the-olympics.

Molik, Bartosz, James J Laskin, Amanda L Golbeck, Andrzej Kosmol, Witold Rekowski, Natalia Morgulec-Adamowicz, Izabela Rutkowska, Jolanta Marszałek, Jan Gajewski, and Miguel-Angel Gomez. "The International Wheelchair Basketball Federation's Classification System: The Participants' Perspective." *Kinesiology* 49, no. 1 (2017): 117–126.

Møller, Verner. "Knud Enemark Jensen's Death During the 1960 Rome Olympics: A Search for Truth?" *Sport in History* 25, no. 3 (2005): 452–471.

Møller, Verner. "One Step Too Far—About WADA's Whereabouts Rule." *International Journal of Sport Policy and Politics* 3, no. 2 (2011): 177–190.

Møller, Verner. "The Road to Hell is Paved with Good Intentions—A Critical Evaluation of WADA's Anti-Doping Campaign." *Performance Enhancement and Health* 4, no. 3–4 (2016): 111–115.

Møller, Verner, and Rasmus Bysted Møller. "Gene Doping: Ethical Perspectives." In *Routledge Handbook of Sport and Exercise System Genetics*, edited by Timothy Lightfoot, Monica J. Hubal, and Stephen M. Roth, 453–462. London: Routledge, 2019.

Møller, Verner, Ivan Waddington, and John Hoberman. *Routledge Handbook of Drugs and Sport*. London: Routledge, 2015.

Monilaw, William J. "The Effects of Training Down in Weight on the Growing Boy and How to Control or Abolish the Practice." *The School Review* 25, no. 5 (1917): 350–360.

Montgomery, Hugh E., Richard Marshall, Harry Hemingway, Saul Myerson, Peter Clarkson, Claire Dollery, Matt Hayward, D. E. Holliman, Mick Jubb, Michael World, Edwin L. Thomas, Audrey E. Brynes, Nasir Saeed, Marina Arnard, Johann D. Bell, Kapa Prasad, Mark Rayson, Philippa J. Talmud, and Steve E. Humphries. "Human Gene for Physical Performance." *Nature* 393 (May 21, 1998): 221–222.

Moore, Eric, and Jo Morrison. "In Defense of Medically Supervised Doping." *Journal of the Philosophy of Sport* (2022): 159–176.

Moore, Jan M., Anna F. Timperio, David A. Crawford, Cate M. Burns, and David Cameron-Smith. "Weight Management and Weight Loss Strategies of Professional Jockeys." *International Journal of Sport Nutrition and Exercise Metabolism* 12, no. 1 (2002): 1–13.

Moore, Keith. "Sexual Identity of Athletes." *Journal of the American Medical Association* 205 (1968): 163–164.

Moran, Colin N., and Guan Wang. "Genetic Limitations to Athletic Performance." In Routledge *Handbook on Biochemistry of Exercise*, edited by Peter M. Tiidus, Rebecca E. K. MacPherson, Paul J. LeBlanc, and Andrea R. Josse, 217–231. London: Routledge, 2020.

Moreau, Henri. "The Genesis of the Metric System and the Work of the International Bureau of Weights and Measures." *Journal of Chemical Education* 30, no. 1 (1953): 3–20.

Mosher, Dana S., Pascale Quignon, Carlos D. Bustamante, Nathan B. Sutter, Cathryn S. Mellersh, Heidi G. Parker, and Elaine A. Ostrander. "A Mutation in the Myostatin Gene Increases Muscle Mass and Enhances Racing Performance in Heterozygote Dogs: E79." *PLoS Genetics* 3, no. 5 (2007), e79.

Mottram, David R., ed. *Drugs in Sport*. London: Routledge, 2012.

Müller, Wolfram, W. Gröschl, R. Müller, and Karl Sudi. "Underweight in Ski Jumping: The Solution to the Problem." *International Journal of Sports Medicine* 27, no. 11 (2005): 926–934.

Müller, Wolfram, Dieter Platzer, and Bernhard Schmölzer. "Dynamics of Human Flight on Skis: Improvement on Safety and Fairness in Ski Jumping." *Journal of Biomechanics* 29, no. 8 (1996): 1061–1068.

Munro, Brenna. "Caster Semenya: Gods and Monsters." *Safundi* 11, no. 4 (2010): 383–396.

Murray, Steven R., and Brian E. Udermann. "Fluid Replacement: A Historical Perspective and Critical Review." *International Sports Journal* (2003): 59–73.

Murray, Thomas H., Karen J. Maschke, and Angela A. Wasunna, eds. *Performance-Enhancing Technologies in Sports: Ethical, Conceptual, and Scientific Issues*. Baltimore: Johns Hopkins University Press, 2009.

Negesa, Annet. "I Cannot Go Back to the Body I Had Before I Was Operated On, But I Can Try to Stop Other Women Going Through What I Did." *Telegraph,* November 18, 2021. https://www.telegraph.co.uk/athletics/2021/11/18/annet-negesa-cannot-go-back-body-had-operated-can-try-stop-women/#:~:text=Cheltenham%20Festival-,Annet%20Neg esa%3A%20'I%20cannot%20go%20back%20to%20the%20body%20 I,going%20through%20what%20I%20did'&text=Nine%20years%20 ago%20sports%20authorities,feminine%2C%E2%80%9D%20or%20q uit%20running.

Neporent, Liz. "Winter Olympic Sports: One Size Doesn't Fit All." *ABC News.* February 11, 2014. https://abcnews.go.com/Sports/winter-olym pic-sports-size-fits/story?id=22447486.

Nevill, Alan M., and Gregory Whyte. "Are There Limits to Running World Records?" *Medicine and Science in Sports and Exercise* 37, no. 10 (2005): 1785–1788.

Nevill, Alan M., Gregory Whyte, Roger L. Holder, and M. Peyrebrune. "Are There Limits to Swimming World Records?" *International Journal of Sports Medicine* 28, no. 12 (2007): 1012–1017.

"No More Figures in Figure Skating." *New York Times,* June 9, 1988. https:// www.nytimes.com/1988/06/09/sports/no-more-figures-in-figure-skat ing.html#:~:text=The%20International%20Skating%20Union%20vo ted,be%20reduced%20in%20dance%20programs

Noakes, Tim. *Waterlogged: The Serious Problem of Overhydration in Endurance Sports.* Champaign, IL: Human Kinetics, 2012.

Novak, Andrew. "Disability Sport in Sub-Saharan Africa: From Economic Underdevelopment to Uneven Empowerment." *Disability and the Global South* 1, no. 1 (2014): 44–63.

Nyong'o, Tavia. "The Unforgivable Transgression of Being Caster Semenya." *Women & Performance: A Journal of Feminist Theory* 20, no. 1 (2010): 95–100.

Obasa, Mojisola, and Pascal Borry. "The Landscape of the 'Spirit of Sport.'" *Journal of Bioethical Inquiry* 16, no. 3 (2019): 443–453.

"Olympic Games." *Time.* August 10, 1936, 28.

"One in Half-Million Chance of Making the Olympics," *Horsetalk,* July 25, 2012. Accessed June 14, 2020, https://www.horsetalk.co.nz/2012/07/25/ one-in-half-million-chance-olympics/.

"One-Legged Man an Athlete." *Morning Oregonian,* May 29, 1908, 7.

Opplinger, Robert A., Ronald D. Harms, Donald E. Herrmann, Cynthia M. Streich, and Randall R. Clark. "The Wisconsin Wrestling Minimal Weight Project: A Model for Wrestling Weight Control." *Medicine and Science in Sports and Exercise* 27, no. 8 (1995): 1220–1224.

Opplinger, Robert A., Suzanne A. Nelson Steen, and James R. Scott. "Weight Loss Practices of College Wrestlers." *International Journal of Sport Nutrition and Exercise Metabolism* 13, no. 1 (2003): 29–46.

Ospina-Betancurt, Jonathan, Eric Vilain, and María José Martinez-Patiño. "The End of Compulsory Gender Verification: Is It Progress for Inclusion of Women in Sports?" *Archives of Sexual Behavior* (September 2021): 1–9.

Ostrander, Elaine A., Heather J. Hudson, and Gary K. Ostrander. "Genetics of Athletic Performance." *Annual Review of Genomics and Human Genetics* 10 (2009): 407–429.

Papadimitriou, Ioannis D., Sarah J. Lockey, Sarah Voisin, Adam J. Herbert, Fleur Garton, Peter J. Houweling, Pawel Cieszczyk, Agnieszka Maciejewska-Skrendo, Marek Sawczuk, Myosotis Massidda, Carla Maria Calò, Irina V. Astratenkova, Anastasia Kouvatsi, Anastasiya M. Druzhevskaya, Macsue Jacques, Ildus I. Ahmetov, Georgina K. Stebbings, Shane Heffernan, Stephen H. Day, Robert Erskine, Charles Pedlar, Courtney Kipps, Kathryn N. North, Alun G. Williams, and Nir Eynon. "No Association Between ACTN3 R577X and ACE I/D Polymorphisms and Endurance Running Times in 698 Caucasian Athletes." *BMC Genomics* 19, no. 1 (2018): 1–9.

"Paralympics: Mallory Weggemann Shocked by Classification Change." *BBC,* August 30, 2012. https://www.bbc.com/sport/disability-sport/19429915.

Parisotto, Robin. *Blood Sports: The Inside Dope on Drugs in Sport.* Melbourne: Hardie Grant, 2006.

Patel, Seema. "Gaps in the Protection of Athletes Gender Rights in Sport—A Regulatory Riddle." *International Sports Law Journal* 21 (2021): 257–275.

Patel, Seema. "Rugby, Concussions and Duty of Care: Why the Game Is Facing Scrutiny." *The Conversation,* June 3, 2021. https://theconversation.com/rugby-concussions-and-duty-of-care-why-the-game-is-facing-scrutiny-161773.

Paul, William D. "Crash Diets and Wrestling." *Journal of the Iowa Medical Society* 56, no. 8 (1966): 835–840.

Pavitt, Michael. "IPC Says Up to 75 Wheelchair Basketball Athletes Should Undergo Reclassification in Time for Tokyo 2020." *Inside the Games,* February 9, 2020. https://www.insidethegames.biz/articles/1090325/ipc-wheelchair-basketball-tokyo-2020.

Peers, Danielle. "Interrogating Disability: The (De)composition of a Recovering Paralympian." *Qualitative Research in Sport, Exercise and Health* 4, no. (2012): 175–188.

Peters, Mary, with Ian Wooldridge. *Mary P.: Autobiography.* London: Stanley Paul, 1974.

Peters, Sam. "England Stars Blocked RFU Concussion Gene-Testing Plan for All Professional Players Due to 'Big Brother' Privacy Fears." *Daily Mail,* November 29, 2014. https://www.dailymail.co.uk/sport/concussion/article-2854458/England-stars-blocked-RFU-concussion-gene-testing-plan-professional-players-Big-Brother-privacy-fears.html.

Pfleegor, Adam G., and Danny Roesenberg. "Deception in Sport: A New Taxonomy of Intra-lusory Guiles." *Journal of the Philosophy of Sport* 41, no. 2 (2014): 209–231.

Pickering, Craig, and John Kiely. "Can Genetic Testing Predict Talent? A Case Study of 5 Elite Athletes." *International Journal of Sports Physiology and Performance* 16, no. 3 (2020): 429–434.

Pickering, Craig, John Kiely, Jozo Grgic, Alejandro Lucia, and Juan Del Coso. "Can Genetic Testing Identify Talent for Sport?" *Genes* 10, no. 972 (2019), 1–13.

Pielke, Roger. "Caster Semenya Ruling: Sports Federation is Flouting Ethics Rules." *Nature.* May 17, 2019. https://www.nature.com/articles/d41 586-019-01606-8.

Pielke, Roger Jr.. *The Edge: The War Against Cheating and Corruption in the Cutthroat World of Elite Sport.* Berkeley, CA: Roaring Forties Press, 2016.

Pielke, Roger, Ross Tucker, and Erik Boye. "Scientific Integrity and the IAAF Testosterone Regulations." *International Sports Law Journal* 19, no. 1 (2019): 18–26.

Pieper, Lindsay Parks. "Gender Regulation: Renée Richards Revisited." *International Journal of the History of Sport* 29, no. 5 (2012): 675–690.

Pieper, Lindsay Parks. "The Medical Examination of Lady Competitors: Sex Control in Skiing, 1967–2000." *International Journal of the History of Sport* 10, no. 9 (2021): 1–19.

Pieper, Lindsay Parks. *Sex Testing: Gender Policing in Women's Sports.* Urbana: University of Illinois Press, 2016.

Pitsiladis, Yannis, Guan Wang, Alain Lacoste, Christian Schneider, Angela D. Smith, Alessia Di Gianfrancesco, and Fabio Pigozzi. "Make Sport Great Again: The Use and Abuse of the Therapeutic Use Exemptions Process." *Current Sports Medicine Reports* 16, no. 3 (2017): 123–125.

"The 'Police Gazette' Champion Belt." *National Police Gazette,* November 1, 1884, 13.

Pope Jr., Harrison G., Ruth I. Wood, Alan Rogol, Fred Nyberg, Larry Bowers, and Shalender Bhasin. "Adverse Health Consequences of Performance-Enhancing Drugs: An Endocrine Society Scientific Statement." *Endocrine Reviews* 35, no. 3 (2014): 341–375.

Posberg, Anna. "Defining 'Woman': A Governmentality Analysis of How Protective Policies Are Created in Elite Women's Sport." *International Review for the Sociology of Sport* (2022): 1–21.

Posthumus, Michael, and Malcolm Collins, eds. *Genetics and Sports.* 2nd ed. New York: Karger, 2016.

Prebish, Rob. *The Solitary Wrestler: Methods for Safe Weight Control* (JBE Online Books, 2006), http://www.jbeonlinebooks.org/wrestling/doc-uments/SolitaryWrestler-Chapter04.pdf.

"Preserving la Difference." *Time,* September 16, 1966, 74.

Press, Joel M., Patricia Dietz Davis, Steven L. Wiesner, Allen Heinemann, Patrick Semik, and Robert G. Addison. "The National Jockey Injury Study: An Analysis of Injuries to Professional Horse-racing Jockeys." *Clinical Journal of Sport Medicine* 5, no. 4 (1995): 236–240.

Prokop, Ludwig. "The Struggle Against Doping and Its History." *Journal of Sports Medicine and Physical Fitness* 10, no. 1 (1970): 45–48.

Puce, Luca, Lucio Marinelli, Ilaria Pallecchi, Laura Mori, and Carlo Trompetto. "Impact of the 2018 World Para Swimming Classification Revision on the Race Results in International Paralympic Swimming Events." *German Journal of Exercise and Sport Research* (2019): 1–13.

Rader, Benjamin G. *American Sports: From the Age of Folk Games to the Age of Televised Sports.* 3rd ed. Englewood Cliffs, NJ: Prentice Hall, 1996.

Rahim, Masouda, Malcolm Collins, and Alison September. "Genes and Musculoskeletal Soft-Tissue Injuries." In *Genetics and Sports,* 2nd ed., edited by Michael Posthumus and Malcolm Collins, 68–91. New York: Karger, 2016.

"Raising Competition Age for Figure Skaters Not Enough to Combat Abusive Coaches, Former Skaters Say." *CBC,* June 8, 2022. https://ca.news.yahoo.com/raising-competition-age-figure-skaters-003721081.html.

"Raising the Age Limit for Skating Would End the Age of the Quad." *USA Today,* February 18, 2022. https://www.usatoday.com/story/sports/olympics/2022/02/18/raising-age-limit-for-skating-would-end-the-age-of-the-quad/49830469/.

Rand, Frederick Rogers. "Olympics for Girls?" *School & Society* 30 (August 1929): 194.

Ransone, Jack, and Brian Hughes. "Body-Weight Fluctuations in Collegiate Wrestlers: Implications of the National Collegiate Athletic Association Weight-Certification Program." *Journal of Athletic Training* 39, no. 2 (2004): 162–165.

Rantanen, Elina, Marja Hietala, Ulf Kristoffersson, Irmgard Nippert, Jörg Schmidtke, Jorge Sequeiros, and Helena Kääriäinen. "What Is Ideal Genetic Counselling? A Survey of Current International Guidelines." *European Journal of Human Genetics* 16, no. 4 (2008): 445–452.

Rasch, Philip J., and Walter Kroll. *What the Research Tells the Coach about Wrestling.* Washington, DC: American Association for Health, Physical Education, and Recreation, 1964.

Rasmussen, Nicolas. *On Speed: From Benzedrine to Adderall.* New York: NYU Press, 2008.

Read, Daniel, James Skinner, Daniel Lock, and Aaron C. T. Smith. *WADA, the World Anti-Doping Agency: A Multi-Level Legitimacy Analysis.* London: Routledge, 2021.

"Records of Polish Girl Sprinter Who Flunked Sex Test Barred." *New York Times,* February 26, 1968, 50.

Reynolds, Allie, and Alireza Hamidian Jahromi. "Transgender Athletes in Sports Competitions: How Policy Measures Can Be More Inclusive and Fairer to All." *Frontiers in Sports and Active Living* 3 (2021). https://doi.org/10.3389/fspor.2021.704178.

Reynolds, Gretchen. "Outlaw DNA." *New York Times,* June 3, 2007. https://www.nytimes.com/2007/06/03/sports/playmagazine/0603play-hot.html.

Ribisi, Paul M., and William G. Herbert. "Effects of Rapid Weight Reduction and Subsequent Rehydration upon the Physical Working Capacity of Wrestlers." *Research Quarterly* 41, no. 4 (1970): 536–541.

Rice, Grantland. "Separate Olympics for Sexes in 1940 Planned." *Los Angeles Times,* August 12, 1936, A9.

Rice, James. *History of the British Turf, from the Earliest Times to the Present Day,* vol. II. London: Sampson Low, Marston, Searle, and Rivington, 1879.

Ritchie, Ian. "The Construction of a Policy: The World Anti-Doping Code's 'Spirit of Sport' Clause." *Performance Enhancement & Health* 2, no. 4 (2013): 194–200.

Rodriguez, Robert G. *The Regulation of Boxing.* Jefferson, NC: McFarland, 2009.

Roenigk, Alyssa. "Karolyi Says Age Limit Would Rob Gymnasts of Golden Opportunity." *ESPN,* August 10, 2008. https://www.espn.com/olympics/summer08/gymnastics/columns/story?id=3527997.

Rosewater, Amy. "Are Officials Redefining the Paralympian?" *ESPN,* July 15, 2015. http://www.espn.com/olympics/story/_/id/13258441/paralympics-officials-redefining-paralympian.

Rottenberg, Simon. "The Baseball Players' Labor Market." *Journal of Political Economy,* 64 (1956): 242–258.

Rous, Henry John. *On the Laws and Practice of Horse Racing.* London: A. H. Baily, 1866.

Ruggie, John G. *"For the Game. For the World."* FIFA and Human Rights. Corporate Responsibility Initiative Report No. 68. Cambridge, MA: Harvard Kennedy School, 2016.

Rusko, Heikki R. "New Aspects of Altitude Training." *American Journal of Sports Medicine* 24, no. 6 (1996): S48–S52.

Ruwuya, Jonathan, Byron Omwando Juma, and Jules Woolf. "Challenges Associated with Implementing Anti-Doping Policy and Programs in Africa." *Frontiers in Sports and Active Living* 4 (December 8, 2022): 966559. doi:10.3389/fspor.2022.966559.

Ryan, Joan. "When Will Child Athletes Be Protected?" *Washington Post,* February 21, 2022, A17.

Saletan, William. "The Beam in Your Eye: If Steroids Are Cheating, Why Isn't LASIK?" *Slate,* April 18, 2005. https://slate.com/technology/2005/04/if-steroids-are-cheating-why-isn-t-lasik.html.

Sammons, Jeffrey. *Beyond the Ring: The Role of Boxing in American Society.* University of Illinois Press, 1988.

Samuel, Sigal. "How Biohackers Are Trying to Upgrade Their Brains, Their Bodies—and Human Nature." *Vox,* November 15, 2019. https://www.vox.com/future-perfect/2019/6/25/18682583/biohacking-transhumanism-human-augmentation-genetic-engineering-crispr.

Savulescu, Julian. "Ten Ethical Flaws in the Caster Semenya Decision on Intersex in Sport." *The Conversation.* May 9, 2019, https://theconversation.com/ten-ethical-flaws-in-the-caster-semenya-decision-on-intersex-in-sport-116448.

Savulescu, Julian. "Why It's Time to Legalize Doping in Athletics." *The Conversation.* August 28, 2015. https://theconversation.com/why-its-time-to-legalise-doping-in-athletics-46514.

Savulescu, Julian, and Nick Bostrom, eds. *Human Enhancement.* New York: Oxford University Press, 2009.

Savulescu, Julian, and Bennett Foddy. "Comment: Genetic Test Available for Sports Performance." *British Journal of Sports Medicine* 39, no. 8 (2005): 472.

Schmölzer, Bernhard, and Wolfram Müller. "The Importance of Being Light: Aerodynamic Forces and Weight in Ski Jumping." *Journal of Biomechanics* 35, no. 8 (2002): 1059–1069.

Schneider, Angela. "Banned from the Tokyo Olympics for Pot? Let the Athletes Decide What Drugs Should Be Allowed." *The Conversation,* July 6, 2021. https://theconversation.com/banned-from-the-tokyo-olympics-for-pot-let-the-athletes-decide-what-drugs-should-be-allowed-163619.

Schneider, Angela J. "Privacy, Confidentiality and Human Rights in Sport." *Sport in Society* 7, no. 3 (2004): 438–456.

Schneider, Angela J., and Theodore Friedmann. *Gene Doping in Sports: The Science and Ethics of Genetically Modified Athletes.* Boston: Elsevier, 2006.

Schuelke, Markus, Kathryn R. Wagner, Leslie E. Stolz, Christoph Hübner, Thomas Riebel, Wolfgang Kömen, Thomas Braun, James F. Tobin, and Se-Jin Lee. "Myostatin Mutation Associated with Gross Muscle Hypertrophy in a Child." *New England Journal of Medicine* 350, no. 26 (2004): 2682–2688.

Schult, Olivia, and Laura Rivard. "Case Study in Genetic Testing for Sports Ability." *Nature,* September 25, 2013. https://www.nature.com/scitable/forums/genetics-generation/case-study-in-genetic-testing-for-sports-107403644/.

Schultz, Jaime. "Caster Semenya and the 'Question of Too': Sex Testing in Elite Women's Sport and the Issue of Advantage." *Quest* 63, no. 2 (2011): 228–243.

Schultz, Jaime. "Good Enough? The 'Wicked' Use of Testosterone for Defining Femaleness in Women's Sport." *Sport in Society* 24, no. 4 (2021): 607–627.

Schultz, Jaime. *Qualifying Times: Points of Change in U.S. Women's Sport.* Urbana: University of Illinois Press, 2014.

Schultz, Jaime. *Women's Sports: What Everyone Needs to Know.* New York: Oxford University Press, 2018.

Schultz, Jaime, W. Lawrence Kenney, and Andrew D. Linden. "Heat-Related Deaths in American Football: An Interdisciplinary Approach." *Sport History Review* 45, no. 2 (2014): 123–144.

Schwarz, Alan. "Baseball's Use of DNA Tests Raises Ethical Issues." *International Herald Tribune,* July 22, 2009.

Schwartz, Hillel. *Never Satisfied: A Cultural History of Diets, Fantasies and Fat.* New York: The Free Press, 1986.

"Semenya 'Maybe Not100pc' a Woman." *ABC News,* September 11, 2009. http://www.abc.net.au/news/2009-09-11/semenya-maybe-not-100pc-a-woman/1424994.

"Sex Test Disqualifies Athlete." *New York Times* September 16, 1967, 28.

Sheard, Kenneth G. "Aspects of Boxing in the Western Civilizing Process." *International Review for the Sociology of Sport* 32, no. 1 (1997): 31–57.

Sherrill, Claudine. "Disability Sport and Classification Theory: A New Era." *Adapted Physical Activity Quarterly* 16, no. 3 (1999): 206–215.

Sherrill. Claudine, ed. *Sport and Disabled Athletes.* Champaign, IL: Human Kinetics, 1986.

Sherrill, Claudine, Carol Adams-Mushett, and Jeffrey Al Jones. "Classification and other Issues in Sports for Blind, Cerebral Palsied, Les Autres, and Amputee Athletes." In *Sport and Disabled Athletes*, edited by Claudine Sherrill, 113–130. Champaign, IL: Human Kinetics, 1986.

Silva, Analiza M., Diana A. Santos, and Catarina N. Matias. "Weight-Sensitive Sports." In *Body Composition: Health and Performance in Exercise and Sport*, edited by Henry C. Lukaski, 233–284. Boca Raton, FL: CRC Press, 2017.

Silverman, Ian. "Ian Silverman Writes Letter to IPC about Classification Issues." *SwimSwam,* July 31, 2016. https://swimswam.com/paralympic-champ-ian-silverman-writes-letter-to-ipc-about-fraud/.

Simpson, Joe Leigh, Arne Ljungqvist, Albert de la Chapelle, Malcolm A. Ferguson-Smith, Myron Genel, Alison S. Carlson, Anke A. Ehrhardt, and Elizabeth Ferris. "Gender Verification in Competitive Sports." *Sports Medicine* 16, no. 5 (1993): 305–315.

Singh, Hameet Shah. "India Athlete Makes Plea for Semenya." *CNN,* September 14, 2009. http://www.cnn.com/2009/WORLD/asiapcf/09/14/Semenya.India.Athlete/index.html.

"Ski Jumpers Are Taking the Light Approach." *Washington Times.* February 9, 2002. https://www.washingtontimes.com/news/2002/feb/9/20020209-035615-1739r/.

Skirstad, Berit. "Gender Verification in Competitive Sport: Turning from Research to Action." In *Values in Sport: Elitism, Nationalism, Gender Equality and the Scientific Manufacture of Winners*, edited by Torbjörn Tännsjö and Claudio Marcello Tamburrini, 116–122. London: Taylor & Francis, 2000.

Slot, Owen. "Apocalypse Now: Fears of Gene Doping Are Realised." *The Times,* February 2, 2006, 78.

Smale, Simon. "New Paralympic Wheelchair Basketball Eligibility Rules Have Ruined Dreams, and Raised Significant Questions." *ABC News,* August 22, 2020. https://www.abc.net.au/news/2020-08-23/paralaympics-wheelchair-basketball-reclassification-ipc/12539302.

Smith, Aaron C. T., and Bob Stewart. "Why the War on Drugs in Sport Will Never Be Won." *Harm Reduction Journal* 12, no. 1 (2015): 1–6.

Smith, Andrew, and Nigel Thomas. "The 'Inclusion' of Elite Athletes with Disabilities in the 2002 Manchester Commonwealth Games: An Exploratory Analysis of British Newspaper Coverage." *Sport, Education and Society* 10, no. 1 (2005): 49–67.

Smith, Stephen. "What Is the Real Cost of Injuries in Professional Sport." *Kitman Labs,* April 23, 2016. https://www.kitmanlabs.com/what-is-the-real-cost-of-injuries-in-professional-sport/.

Sönksen, Peter H., L. Dawn Bavington, Tan Boehning, David Cowan, Nishan Guha, Richard Holt, Katrina Karkazis, Malcolm Andrew Ferguson-Smith, Jovan Mircetic, and Dankmar Böhning. "Hyperandrogenism Controversy in Elite Women's Sport: An Examination and Critique of Recent Evidence," *British Journal of Sports Medicine* 52, no. 23 (2018): 1481–1482.

Sönksen, Peter, Malcolm A. Ferguson-Smith, L. Dawn Bavington, Richard I. G. Holt, David A. Cowan, Don H. Catlin, Bruce Kidd, Georgiann Davis, Paul Davis, Lisa Edwards, and Anne Tamar-Mattis. "Medical and Ethical Concerns Regarding Women with Hyperandrogenism and Elite Sport." *Journal of Clinical Endocrinology and Metabolism* 100, no. 3 (2015): 825–827.

Stanford University. "Are You Elite?" ELITE, https://elite.stanford.edu/.

Steadward, Robert D. "Integration and Sport in the Paralympic Movement." *Sport Science Review* 5, no. 1 (1996): 26–41.

Stearns, Peter N. *Fat History: Bodies and Beauty in the Modern West.* New York: NYU Press, 2002.

Steen, Suzanne Nelson, and Kelly D. Brownell. "Patterns of Weight Loss and Regain in Wrestlers: Has the Tradition Changed?" *Medicine and Science in Sports and Exercise* 22, no. 6 (1990): 762–768.

Sterling, Jennifer J., and Mary G. McDonald, eds. *Sports, Society, and Technology: Bodies, Practices, and Knowledge Production.* Gate East, Singapore: Palgrave Macmillan, 2020.

Stewart, Lauryn, P. O'Halloran, Jennifer Oates, E. Sherry, and Ryan Storr. "Developing Trans-athlete Policy in Australian National Sport Organizations." *International Journal of Sport Policy and Politics* 13, no. 4 (2021): 565–585.

Strohkendl, Horst. *The 50th Anniversary of Wheelchair Basketball.* New York: Waxman, 1996.

Storr, Ryan, Madeleine Pape, and Sheree Bekker. "A Win for Transgender Athletes and Athletes with Sex Variations." *The Conversation,* November 18, 2021. https://theconversation.com/a-win-for-transgender-athle tes-and-athletes-with-sex-variations-the-olympics-shifts-away-from-testosterone-tests-and-toward-human-rights-172045?utm_ source=facebook&utm_medium=bylinefacebookbutton&fbclid= IwAR36WYmbOdtU7UBwxj3b1w9ZEkgJyuXKClKFXLiBn14co 2rDYT5k-8SmG-Q.

Stutfield, G. Herbert. "Handicaps." *National Review* 31, no. 183 (May 1898): 386–402.

Sudai, Maayan. "The Testosterone Rule—Constructing Fairness in Professional Sport." *Journal of Law and Biosciences* 4, no. 1 (2017): 181–193.

Sullivan, Robert. "Triumphs Tainted with Blood." *Sports Illustrated,* January 21, 1985. https://www.si.com/vault/1985/01/21/546256/triumphs-tain ted-with-blood.

Sundgot-Borgen, Jorunn, and Ina Garthe. "Elite Athletes in Aesthetic and Olympic Weight-class Sports and the Challenge of Body Weight and

Body Compositions." *Journal of Sport Sciences* 29, supplement 1 (2011): S101–S114.

Svensson, Eric C., Hugh B. Black, Debra L. Dugger, Sandeep K. Tripathy, Eugene Goldwasser, Zengping Hao, Lien Chu, and Jeffrey M. Leiden. "Long-Term Erythropoietin Expression in Rodents and Non-Human Primates Following Intramuscular Injection of a Replication-Defective Adenoviral Vector." *Human Gene Therapy* 8, no. 15 (1997): 1797–1806.

Swartz, Leslie, Jason Bantjes, Divan Rall, Suzanne Ferreira, Cheri Blauwet, and Wayne Derman. "'A More Equitable Society': The Politics of Global Fairness in Paralympic Sport." *PLOS One* 11, no. 12 (2016): e0167481. doi: 10.1371/journal/pone.0167481.

Synovitz, Ron, and Zamira Eshanova. "Uzbekistan Is Using Genetic Testing to Find Future Olympians." *The Atlantic,* February 4, 2014. https://www.theatlantic.com/international/archive/2014/02/uzbekistan-is-using-genetic-testing-to-find-future-olympians/283001/.

Szczerba, Robert J. "Mixed Martial Arts and the Evolution of John McCain." *Forbes,* April 3, 2014. https://www.forbes.com/sites/robertszczerba/2014/04/03/mixed-martial-arts-and-the-evolution-of-john-mccain/?sh=7ff53dc42d59.

Taboga, Paolo, Owen N. Beck, and Alena M. Grabowski. "Prosthetic Shape, but Not Stiffness or Height, Affects the Maximum Speed of Sprinters with Bilateral Transtibial Amputations." *PloS One* 15, no. 2 (2020): e0229035–e0229035.

Tamburrini, Claudio. "Are Doping Sanctions Justified? A Moral Relativistic View." *Sport in Society* 9, no. 2 (2006): 199–211.

Tamburrini, Claudio M. "What's Wrong with Doping"? In *Values in Sport: Elitism, Nationalism, Gender Equality and the Scientific Manufacture of Winners,* edited by Torbjörn Tännsjö and Claudio Tamburrini, 210–226. London: E&FN Spon, 2000.

Tannenbaum, Cara, and Sheree Bekker. "Sex, Gender, and Sports." *British Medical Journal* 364 (2019): l1120.

Tännsjö, Torbjörn. "Medical Enhancement and the Ethos of Elite Sport." In *Human Enhancement,* edited by Julian Savulescu and Nick Bostrom, 315–326. New York: Oxford University Press.

Tännsjö, Torbjörn, and Claudio Tamburrini, eds. *Values in Sport: Elitism, Nationalism, Gender Equality and the Scientific Manufacture of Winners.* London: E&FN Spon, 2000.

Tebbutt, Clare. "The Spectre of the 'Man-woman Athlete': Mark Weston, Zdenek Koubek, the 1936 Olympics and the Uncertainty of Sex." *Women's History Review* 24, no. 5 (2015): 721–738.

Teetzel, Sarah. "Minimum and Maximum Age Limits for Competing in the Olympic Games." In *Proceedings: International Symposium for Olympic Research*, 340–347. London, Ontario: International Centre for Olympic Studies, 2010.

Teetzel, Sarah. "Philosophic Perspectives on Doping Sanctions and Young Athletes." *Frontiers in Sports and Active Living* (2022): https://doi.org/10.3389/fspor.2022.841033.

"Testers Treat Us Like Criminals, Says Nadal." *The Guardian,* February 12, 2009. https://www.theguardian.com/sport/2009/feb/12/rafael-nadal-drugs-test-criminals-olympic-andy-murray.

Tharabenjasin, Phuntila, Noel Pabalan, and Hamdi Jarjanazi. "Association of the ACTN3 R577X (rs1815739) Polymorphism with Elite Power Sports: A Meta-Analysis." *PloS one* 14, no. 5 (2019): e0217390.

Thevis, Mario, Hans Geyer, Andreas Thomas, and Wilhelm Schänzer. "Trafficking of Drug Candidates Relevant for Sports Drug Testing: Detection of Non-approved Therapeutics Categorized as Anabolic and Gene Doping Agents in Products Distributed via the Internet." *Drug Testing and Analysis* 3, no. 5 (2011): 331–336.

Thibault, Valérie, Marion Guillaume, Geoffroy Berthelot, Nour El Helou, Karine Schaal, Laurent Quinquis, Hala Nassif, Muriel Tafflet, Sylvie Escolano, Olivier Hermine, and Jean-François Toussaint. "Women and Men in Sport Performance: The Gender Gap Has Not Evolved Since 1983." *Journal of Sports Science and Medicine* 9, no. 2 (2010): 214–223.

Thompson, Christopher S. *The Tour de France: A Cultural History*. Berkeley: University of California Press, 2006.

Tiidus, Peter M., Rebecca E. K. MacPherson, Paul J. LeBlanc, and Andrea R. Josse, eds. *Routledge Handbook on Biochemistry of Exercise*. London: Routledge, 2020.

Tipton, Charles M., and Tse-Kia Tcheng. "Iowa Wrestling Study: Weight Loss in High School Students." *JAMA* 214, no. 7 (1970): 1269–1274.

Tipton, Charles M., Tse-Kia Tcheng, and W. D. Paul. "Evaluation of the Hall Method for Determining Minimum Wrestling Weights." *Journal of the Iowa Medical Society* 59 (1969): 571–574.

Todd, Jan, and Terry Todd. "Reflections on the 'Parallel Federation Solution' to the Problem of Drug Use in Sport." In *Performance-Enhancing Technologies in Sports: Ethical, Conceptual, and Scientific Issues*, edited by Thomas H. Murray, Karen J. Maschke, and Angela A. Wasunna, 65–76. Baltimore: Johns Hopkins University Press, 2009.

Todd, Jan, and Terry Todd. "Significant Events in the History of Drug Testing and the Olympic Movement: 1960–1999." In *Doping in Elite Sport: Politics of Drugs in the Olympic Movement,* edited by Wayne Wilson and Edard Derse, 65–128. Champaign, IL: Human Kinetics, 2001.

Todd, Terry. "Anabolic Steroids: The Gremlins of Sport." *Journal of Sport History* 14, no. 1 (1987): 87–107.

Tolich, Martin, and Martha Bell. "The Commodification of Jockeys' Working Bodies: Anorexia or Work Discipline." In *A Global Racecourse: Work, Culture, and Horse Sports,* edited by Chris McConville, 101–133. Melville: Australian Society for Sport History, 2008.

Tolleneer, Jan Sigrid Sterckx, Pieter Bonte, eds. *Athletic Enhancement, Human Nature, and Ethics.* Dordrecht: Springer, 2013.

"Trainers 'Do No Use Female Jockeys' Despite Findings of New Study." *BBC,* January 30, 2018. https://www.bbc.com/sport/horse-racing/42870971.

Tranter, Neil. *Sport, Economy and Society in Britain 1750–1914.* Cambridge: Cambridge University Press, 1998.

Tuakli-Wosornu, Yetsa A., Sandi L. Kirby, Anne Tivas, and Daniel Rhind. "The Journey to Reporting Child Protection Violations in Sport: Stakeholder Perspectives." *Frontiers in Psychology* 13 (2022). https://doi.org/10.3389/fpsyg.2022.907247 https://www.frontiersin.org/articles/10.3389/fpsyg.2022.907247/full.

Tucker, Ross. "Hyperandrogenism and Women vs Women vs Men in Sport: A Q&A with Joanna Harper." *Science of Sport,* May 23, 2016. https://sportsscientists.com/2016/05/hyperandrogenism-women-vs-women-vs-men-sport-qa-joanna-harper/.

Tuttle, W.W. "The Effect of Weight Loss by Dehydration and the Withholding of Food on the Physiologic Responses of Wrestlers." *Research Quarterly* 14 (1943): 158–166.

Tweedy, Sean M. "Taxonomic Theory and the ICF: Foundations for a Unified Disability Athletics Classification." *Adapted Physical Activity Quarterly* 19, no. 2 (2002): 220–237.

Tweedy, Sean M., Emma M. Beckman, and Mark J. Connick. "Paralympic Classification: Conceptual Basis, Current Methods, and Research Update." *PM&R* 6, no. 8 (2014): S11–S17.

Tweedy, Sean M., Mark J. Connick, Emma M. Beckman. "Applying Scientific Principles to Enhance Paralympic Classification Now and in the Future: A Research Primer for Rehabilitation Specialists." *Physical Medicine and Rehabilitation Clinics of North America* 29 (2018): 313–332.

Tweedy, Sean M., and Yves C. Vanlandewijck. "International Paralympic Committee Position Stand—Background and Scientific Principles of

Classification in Paralympic Sport." *British Journal of Sports Medicine* 45 (2011): 259–269.

UK Sport. "UK Sport Statement on Funding." November 25, 2004. https:// www.uksport.gov.uk/news/2004/11/25/uk-sport-statement-on-funding.

Ulrich, Rolf, Harrison G. Pope, Jr., Léa Cléret, Andrea Petróczi, Tamás Nepusz, Jay Schaffer, Gen Kanayama, R. Dawn Comstock, Perikles Simon. "Doping in Two Elite Athletics Competitions Assessed by Randomized-Response Surveys." *Sports Medicine* 48, no. 1 (2018): 211–219.

Ungerleider, Steven, and Bill Bradley. *Faust's Gold: Inside the East German Doping Machine.* New York: Macmillan, 2001.

United Nations. *Convention on the Rights of the Child.* 1989. https://www. ohchr.org/en/instruments-mechanisms/instruments/convention-rig hts-child.

United Nations. "Factsheet on Persons with Disabilities." Accessed June 25, 2021. https://www.un.org/development/desa/disabilities/resources/ factsheet-on-persons-with-disabilities.html.

Vamplew, Wray. "Playing with the Rules: Influences on the Development of Regulation in Sport." *International Journal of the History of Sport* 24, no. 7 (2007): 843–871.

Vamplew, Wray. "Reduced Horse Power: The Jockey Club and the Regulation of British Horseracing." *Entertainment Law* 2, no. 3 (2003): 94–95.

Vamplew, Wray. "Still Crazy After All Those Years: Continuity in a Changing Labour Market for Professional Jockeys." *Sport in Society* 19, no. 3 (2016): 378–399.

Vamplew, Wray. *The Turf: A Social and Economic History of Horse Racing.* London: Allen Lane, 1976.

Van Bottenburg, Maarten, Arnout Geeraert, and Oliver de Hon. "The World Anti-Doping Agency: Guardian of Elite Sport's Credibility." In *Guardians of Public Value: How Public Organizations Become and Remain Institutions,* edited by Arjen Boin, Lauren A. Fahy, and Paul 't Hart. Cham, 185–210. Switzerland: Palgrave Macmillan, 2021.

Van Dornick, Kirsti, and Nancy L. I. Spencer. "What's in A Sport Class? The Classification Experiences of Paraswimmers." *Adapted Physical Activity Quarterly* 37, no. 1 (2020): 1–19.

Van Mill, David. "Why Are We So Opposed to Performance-Enhancing Drugs in Sport?" *The Conversation,* August 27, 2015. http://thec onversation.com/why-are-we-so-opposed-to-performance-enhanc ing-drugs-in-sport-46528.

Vanlandewijck, Yves C., and Rudi J. Chappel. "Integration and Classification Issues in Competitive Sports for Athletes with Disabilities." *Sport Science Review* 5, no. 5 (1996): 65–88.

Varley, Ian, David C. Hughes, Julie P. Greeves, Trent Stellingwerff, Craig Ranson, William D. Fraser, and Craig Sale. "The Association of Novel Polymorphisms with Stress Fracture Injury in Elite Athletes: Further Insights from the SFEA Cohort." *Journal of Science and Medicine in Sport* 21, no. 6 (2018): 564–568.

Veber, Michael. "The Coercion Argument Against Performance-Enhancing Drugs." *Journal of the Philosophy of Sport* 41, no. 2 (2014): 267–277.

Velija, Philippa, and Leah Flynn. "'Their Bottoms are the Wrong Shape': Female Jockeys and the Theory of Established Outsider Relations." *Sociology of Sport Journal* 27, no. 3 (2010): 301–315.

Venezia, Andrew C., and Stephen M. Roth. "The Scientific and Ethical Challenges of Using Genetics Information to Predict Sport Performance." In *Routledge Handbook of Sport and Exercise Systems Genetics*, edited by J. Timothy Lightfoot, Monica J. Hubal, and Stephen M. Roth, 442–452. New York: Routledge, 2019.

Vines, Gail. "Last Olympics for the Sex Test?" *New Scientist* 132, no. 1828 (1992): 39–42.

Virmavirta, Mikko, and Juha Kivekäs. "Is It Still Important to Be Light in Ski Jumping?" *Sports Biomechanics* 20, no. 4 (2021): 407–418.

Vlahovich, Nicole, Peter A. Fricker, Matthew A. Brown, and David Hughes. "Ethics of Genetic Testing and Research in Sport: A Position Statement from the Australian Institute of Sport." *British Journal of Sports Medicine* 51, no. 1 (2017): 5–11.

Voet, Willy. *Breaking the Chain, Drugs and Cycling: The True Story*. Trans. William Fotheringham. London: Yellow Jersey Press, 2001.

Von Hippel, Paul T., Caroline G. Rutherford, and Katherine M. Keyes. "Gender and Weight Among Thoroughbred Jockeys: Underrepresented Women and Underweight Men." *Socius* 3 (2017): 1–7.

Voss, Natalie. "'They Just Want to Ride': Small Changes in Scale of Weights Have Big Impact on Jockeys' Health." *Paulick Report,* February 8, 2017. https://www.paulickreport.com/news/ray-s-paddock/just-want-ride-small-changes-scale-weights-big-impact-jockeys-health/;

Vourvoulias, Bill. "In Kentucky Derby, Race Against Weight Takes Jockeys to a Dark Side." *Fox News,* May 2, 2014. http://www.foxnews.com/sports/2014/05/02/in-kentucky-derby-race-against-weight-takes-jockeys-to-dark-side.html.

Waddington, Ivan. "'A Prison of Measured Time'? A Sociologist Looks at the WADA Whereabouts System." In *Doping and Anti-Doping Policy in Sport: Ethical, Legal and Social Perspectives*, edited by Mike McNamee and Verner Møller, 183–199. London: Taylor & Francis, 2011.

Waddington, Ivan. "Theorising Unintended Consequences of Anti-Doping Policy." *Performance Enhancement and Health* 4, no. 3–4 (2016): 80–87.

Waddington, Ivan, and Verner Møller. "WADA at Twenty: Old Problems and Old Thinking?" *International Journal of Sport Policy and Politics* (2019). doi: 10.1080/19406940.2019.1581645.

Wahlert, Lance, and Autumn Fiester. "Gender Transports: Privileging the 'Natural' in Gender Testing Debates for Intersex and Transgender Athletes." *American Journal of Bioethics* 12, no. 7 (2012): 19–21.

Ware, Mark A., Dennis Jensen, Amy Barrette, Alan Vernec, and Wayne Derman. "Cannabis and the Health and Performance of the Elite Athlete." *Clinical Journal of Sport Medicine* 28, no. 5 (2018): 480–484.

Webborn, Nick, Alun Williams, Mike McNamee, Claude Bouchard, Yannis Pitsiladis, Ildus Ahmetov, Euan Ashley, Nuala Byrne, Silvia Camporesi, Malcolm Collins, Paul Dijkstra, Nir Enyon, Noriyuki Fuku, Fleur C. Garton, Nils Hoppe, Søren Holm, Jane Kaye, Vassilis Klissouras, Alejandro Lucia, Kamiel Maase, Colin Moran, Kathryn N. North, Fabio Pigozzi, and Guan Wang. "Direct-to-Consumer Genetic Testing for Predicting Sports Performance and Talent Identification: Consensus Statement." *British Journal of Sports Medicine* 49, no. 23 (2015): 1486–1491.

Wenner, Melinda. "How to Be Popular during the Olympics: Be H. Lee Sweeney, Gene Doping Expert." *Scientific American*. August 15, 2008. https://www.scientificamerican.com/article/olympics-gene-doping-expert/.

Weyand, Peter G., Lance C. Brooks, Sunil Prajapati, Emily L. McClelland, S. K. Hatcher, Quinn M. Callier, and Matthew W. Bundle. "Artificially Long Legs Directly Enhance Long Sprint Running Performance." *Royal Society Open Science* 9, no. 8 (2022): 220397.

Weyand, Peter G., Matthew W. Bundle, Craig P. McGowan, Alena Grabowski, Mary Beth Brown, Rodger Kram, and Hugh Herr. "The Fastest Runner on Artificial Legs: Different Limbs, Similar Function?" *Journal of Applied Physiology* 107, no. 3 (2009): 903–911.

Wharton, David. "Paralyzed U.S. Swimmer Banned from Paralympic World Championships." *Los Angeles Times*, August 15, 2013. https://www.latimes.com/sports/la-xpm-2013-aug-15-la-sp-sn-paralympic-ban-20130815-story.html.

Whitelaw, Ian. *A Measure of All Things: The Story of Man and Measurement.* New York: St. Martin's Press, 2007.

Wilber, Randall L. "Application of Altitude/Hypoxic Training by Elite Athlete." *Journal of Human Sport and Exercise* 6, no. 2 (2011): 271–286.

Williams, J. G. P. *Sports Medicine.* London: Edward Arnold, 1962.

Wilner, Jon. "Can Superhuman Athletes Provide Genetic Clues on Heart Health?" *Mercury News,* October 29, 2017. https://www.mercurynews.com/2017/10/29/4851089/.

Wilson, George, Jerry Hill, Daniel Martin, James P. Morton, and Graeme L. Close. "GB Apprentice Jockeys Do Not Have the Body Composition to Make Current Minimum Race Weights: Is It Time to Change the Weights or Change the Jockeys?" *International Journal of Sport Nutrition and Exercise Metabolism* 1, no. 2 (2020): 101–104.

Wilson, Wayne, and Edard Derse, eds. *Doping in Elite Sport: Politics of Drugs in the Olympic Movement.* Champaign, IL: Human Kinetics, 2001.

World Anti-Doping Agency. *2020 Anti-Doping Testing Figures Report.* 2020. https://www.wada-ama.org/sites/default/files/2022-01/2020_anti-doping_testing_figures_en.pdf.

World Anti-Doping Agency. *An Athlete's Guide to the Significant Changes in the 2021 Code.* 2021. https://www.athleticsintegrity.org/downloads/pdfs/know-the-rules/en/Athlete-Guide-2021-Code_English_LIVE.pdf.

World Anti-Doping Agency. *Athlete Whereabouts At-a-Glance.* 2021. https://www.wada-ama.org/sites/default/files/2022-03/At_a_Glance_Whereabouts_English_Live_2021.pdf.

World Anti-Doping Agency. "Frequently Asked Questions." Accessed December 22, 2023. https://www.wada-ama.org/sites/default/files/resources/files/athlete_central_faq_final_en_0.pdf.

World Anti-Doping Agency. *Prohibited List, 2021.* Accessed December 22, 2023. https://www.wada-ama.org/sites/default/files/resources/files/2021list_en.pdf.

World Anti-Doping Agency. *Therapeutic Use Exemptions.* 2021. https://www.wada-ama.org/sites/default/files/resources/files/international_standard_istue_-_2021.pdf.

World Anti-Doping Agency. *TUE Physician Guidelines, Medical Information to Support the Decisions of TUE Committees: Female-to-Male (FtM) Transsexual Athletes.* March 2016. https://www.wada-ama.org.

World Anti-Doping Agency. *TUEC Guidelines: Medical Information to Support the Decisions of TUE Committees—Transgender Athletes.* 2017. https://www.wada-ama.org/sites/default/files/resources/files/tuec_transgender_version1.0.pdf.

World Anti-Doping Agency. "WADA Appeals Case of Russian Olympic Committee Figure Skater to Court of Arbitration for Sport." February 21, 2023. https://www.wada-ama.org/en/news/wada-appeals-case-russian-olympic-committee-figure-skater-court-arbitration-sport.

World Anti-Doping Agency. "WADA Ethics Panel: Guiding Values in Sport and Anti-Doping." 2017. https://www.wada-ama.org/sites/default/files/resources/files/wada_ethicspanel_setofnorms_oct2017_en.pdf

World Anti-Doping Agency. "WADA Launches 'Speak Up!'—A Secure Digital Platform to Report Doping Violations." March 9, 2017. https://www.wada-ama.org/en/news/wada-launches-speak-secure-digital-platform-report-doping-violations.

World Anti-Doping Agency. "WADA Note on Artificially Induced Hypoxic Conditions." 2006. http:// www.mcst.go.kr/servlets/eduport/front/upload/UplDownloadFile?pFileName=Note%20on%20Hypoxia%20May%202006.pdf&pRealName=F362307.pdf&pPath=0404150000.

World Anti-Doping Agency. "WADA Statement on Court of Arbitration Decision to Declare Russian Anti-Doping Agency as Non-compliant." December 17, 2020. https://www.wada-ama.org/en/news/wada-statement-court-arbitration-decision-declare-russian-anti-doping-agency-non-compliant.

World Anti-Doping Agency. "WADA Statement on Russian Anti-Doping Agency Finding of 'No Fault or Negligence' in Case of ROC Figure Skater." January 13, 2023. https://www.wada-ama.org/en/news/wada-statement-russian-anti-doping-agency-finding-no-fault-or-negligence-case-roc-figure.

World Anti-Doping Agency. *World Anti-Doping Code, International Standard, Prohibited List, 2024.* Accessed December 22, 2023. https://www.wada-ama.org/sites/default/files/2023-09/2024list_en_final_22_september_2023.pdf.

World Athletics. "World Athletics Council Decides on Russia, Belarus and Female Eligibility." March 23, 2023. https://worldathletics.org/news/press-releases/council-meeting-march-2023-russia-belarus-female-eligibility.

World Bank. "World Bank Country and Lending Groups." Accessed September 16, 2021. https://datahelpdesk.worldbank.org/knowledgebase/articles/906519-world-bank-country-and-lending-groups.

World Health Organization. *World Report on Disability.* Geneva: WHO Press, 2011.

"World Para Athletics Announces Classification Changes." *Athletics Weekly,* October 26, 2017. https://athleticsweekly.com/athletics-news/world-para-athletics-announces-classification-changes-69600/.

World Para Swimming. "Classification in Para Swimming." Accessed December 22, 2023. https://www.paralympic.org/swimming/classification.

World Para Swimming. "World Para Swimming to Introduce Revised Classification Rules and Regulations from 2018." September 29, 2017. https://www.paralympic.org/news/world-para-swimming-introduce-revised-classification-rules-and-regulations-2018.

World Para Winter Sports. "The IPC Provide an Update on Alleged Cases of Intentional Misrepresentation." August 11, 2016. https://www.paralympic.org/news/ipc-provide-update-alleged-cases-intentional-misrepresentation.

Worley, Kristen, and Johanna Schneller. *Woman Enough: How a Boy Became a Woman and Changed the World of Sport.* Toronto: Vintage, 2020.

Yang, Nan, Daniel G. MacArthur, Jason P. Gulbin, Allan G. Hahn, Alan H. Beggs, Simon Easteal, and Kathryn North. "ACTN3 Genotype is Associated with Human Elite Athletic Performance." *American Journal of Human Genetics* 73, no. 3 (2003): 627–631.

Yesalis, Charles E., and Michael S. Bahrke. "History of Doping in Sport." *International Sports Studies* 24, no. 1 (2002): 42–76.

Young, Kevin. "Violence, Risk, and Liability in Male Sports Culture." *Sociology of Sport Journal* 10, no. 4 (1993): 373–396.

Zayner, Josiah. "True Story: I Injected Myself with a CRISPR Genetic Enhancement." *The Antisense,* November 3, 2018. http://theantisense.com/2018/11/13/true-story-i-injected-myself-with-a-crispr-genetic-enhancement/.

Index

For the benefit of digital users, indexed terms that span two pages (e.g., 52–53) may, on occasion, appear on only one of those pages.

Tables and figures are indicated by *t* and *f* following the page number